The Princeton Review®

Cracking the

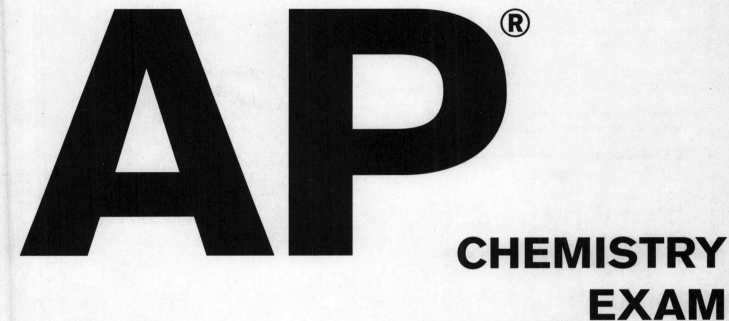

AP®

CHEMISTRY EXAM

2014 Edition

Paul Foglino

PrincetonReview.com

Random House, Inc. New York

The Princeton Review, Inc.
111 Speen Street, Suite 550
Framingham, MA 01701
E-mail: editorialsupport@review.com

ISBN: 978-0-307-94620-1
ISSN: 1092-0102

AP and Advanced Placement Program are registered trademarks of the
College Board, which does not sponsor or endorse this product.

The Princeton Review is not affiliated with Princeton University.

Editor: Calvin S. Cato
Production Editor: Harmony Quiroz
Production Artists: Gabe Berlin, Brandi Valenza, and Deborah Silvestrini

2014 Edition

Editorial
Rob Franek, Senior VP, Publisher
Mary Beth Garrick, Director of Production
Selena Coppock, Senior Editor
Calvin Cato, Editor
Kristen O'Toole, Editor
Meave Shelton, Editor
Alyssa Wolff, Editorial Assistant

Random House Publishing Team
Tom Russell, Publisher
Nicole Benhabib, Publishing Director
Ellen L. Reed, Production Manager
Alison Stoltzfus, Managing Editor

Acknowledgments

The author would like to thank John Katzman for his oversight on this project and Rebecca Lessem for her work editing the first edition of this title. Thanks to Rachel Warren, whose guidance and tireless effort helped to convert chemistry to English. Thanks to Eric Payne for his rigorous attention to detail. Thanks to Tom Meltzer for his advice and Libby O'Connor for her patience. Additional thanks are in order to Robbie Korin, James Karb, and Chris Volpe.

The Princeton Review would like to give special thanks to James Patrick Abulencia for his hard work in updating AP Chemistry to in preparation for the changes to the 2014 edition of the AP Chemistry Exam.

Contents

Part I
Using This Book to Improve Your AP Score

PREVIEW ACTIVITY: YOUR KNOWLEDGE, YOUR EXPECTATIONS

Your route to a high score on the AP Chemistry Exam depends a lot on how you plan to use this book. Respond to the following questions.

1. Rate your level of confidence about your knowledge of the content tested by the Chemistry AP Exam:

   A. Very confident—I know it all
   B. I'm pretty confident, but there are topics for which I could use help
   C. Not confident—I need quite a bit of support
   D. I'm not sure

2. Circle your goal score for the AP Chemistry Exam.

   5   4   3   2   1   I'm not sure yet

3. What do you expect to learn from this book? Circle all that apply to you.

   A. A general overview of the test and what to expect
   B. Strategies for how to approach the test
   C. The content tested by this exam
   D. I'm not sure yet

YOUR GUIDE TO USING THIS BOOK

This book is organized to provide as much—or as little—support as you need, so you can use this book in whatever way will be most helpful to improving your score on the AP Chemistry Exam.

- The remainder of **Part One** will provide guidance on how to use this book and help you determine your strengths and weaknesses

- **Part Two** of this book will:

  - provide information about the structure, scoring, and content of the Chemistry Exam
  - help you to make a study plan
  - point you towards additional resources

- **Part Three** of this book will explore various strategies:

  - how to attack multiple choice questions
  - how to write high scoring free-response answers
  - how to manage your time to maximize the number of points available to you

- **Part Four** of this book covers the content you need for your exam.

- **Part Five** of this book contains practice tests.

You may choose the use some parts of this book over others, or you may work through the entire book. This will depend on your needs and how much time you have. Let's now look how to make this determination.

HOW TO BEGIN

1. **Take a Test**

    Before you can decide how to use this book, you need to take a practice test. Doing so will give you insight into your strengths and weaknesses, and the test will also help you make an effective study plan. If you're feeling test-phobic, remind yourself that a practice test is a tool for diagnosing yourself—it's not how well you do that matters but how you use information gleaned from your performance to guide your preparation.

    So, before you read further, take AP Chemistry Practice Test 1 starting at page 265 of this book. Be sure to do so in one sitting, following the instructions that appear before the test.

2. **Check Your Answers**

    Using the answer key on page 291, count how many multiple choice questions you got right and how many you missed. Don't worry about the explanations for now, and don't worry about why you missed questions. We'll get to that soon.

3. **Reflect on the Test**

    After you take your first test, respond to the following questions:

    - How much time did you spend on the multiple choice questions?

    - How much time did you spend on each long form free response question? What about each short form free response question?

    - How many multiple choice questions did you miss?

    - Do you feel you had the knowledge to address the subject matter of the free response questions?

    - Do you feel you wrote well organized, thoughtful answers to the free response questions?

- Circle the content areas that were most challenging for you and draw a line through the ones in which you felt confident/did well.

  - **Big Idea #1:** Atoms, Elements, and the Building Blocks of Matter

  - **Big Idea #2:** Chemical and Physical Properties of Matter

  - **Big Idea #3:** Chemical Reactions, Energy Changes, and Redox Reactions

  - **Big Idea #4:** Chemical Reactions and their Rates

  - **Big Idea #5:** Laws of Thermodynamics and Changes in Matter

  - **Big Idea #6:** Equilibrium, Acids and Bases, Titrations and Solubility

4. **Read Part Two of this Book and Complete the Self-Evaluation**

   As discussed in the Goals section above, Part Two will provide information on how the test is structured and scored. It will also set out areas of content that are tested.

   As you read Part Two, re-evaluate your answers to the questions above. At the end of Part Two, you will revisit and refine the questions you answer above. You will then be able to make a study plan, based on your needs and time available, that will allow you to use this book most effectively.

5. **Engage with Parts Three and Four as Needed**

   Notice the word *engage*. You'll get more out of this book if you use it intentionally than if you read it passively, hoping for an improved score through osmosis.

   Strategy chapters will help you think about your approach to the question types on this exam. Part Three will open with a reminder to think about how you approach questions now and then close with a reflection section asking you to think about how/whether you will change your approach in the future.

   Content chapters are designed to provide a review of the content tested on the AP Chemistry Exam, including the level of detail you need to know and how the content is tested. You will have the opportunity to assess your mastery of the content of each chapter through test-appropriate questions and a reflection section.

6. **Take Test 2 and Assess Your Performance**

Once you feel you have developed the strategies you need and gained the knowledge you lacked, you should take Test 2, which starts at page 323 of this book. You should do so in one sitting, following the instructions at the beginning of the test.

When you are done, check your answers to the multiple choice sections. See if a teacher will read your answers to the free response questions and provide feedback.

Once you have taken the test, reflect on what areas you still need to work on, and revisit the chapters in this book that address those deficiencies. Through this type of reflection and engagement, you will continue to improve.

7. **Keep Working**

After you have revisited certain chapters in this book, continue the process of testing, reflection, and engaging with the second test. Each time, consider what additional work you need to do and how you will change your strategic approach to different parts of the test.

Part II
About the
AP Chemistry
Exam

THE STRUCTURE OF THE AP CHEMISTRY EXAM

The AP Chemistry Exam is a three-hour-long, two-section test that attempts to cover the material you would learn in a college first-year chemistry course. The first section is a 90-minute 60 question multiple-choice section. The second section is also 90 minutes long and consists of two parts: 3 long-form free response questions and 4 short-form free response questions.

The multiple choice section is scored by a computer and the free-response questions are scored by a committee of high school and college teachers. The free-response questions are graded according to a standard set at the beginning of the grading period by the chief faculty consultants. Inevitably, the grading of Section II is never as consistent or accurate as the grading of Section I.

Please note that the exam is changing for May 2014 and the scoring rubric has not been fully established as of the time of this printing. To receive an accurate breakdown of the approximate raw score you would need in order to receive a 3, 4, or a 5, please check AP Central's Advances in AP website: http://advancesinap. collegeboard.org/science/chemistry.

OVERVIEW OF CONTENT TOPICS

The concepts of the AP Chemistry exam are broken down into six major themes defined by the College Board as the Big Ideas. Rather than learning multiple disparate topics (as has been done in the past), these Big Ideas interconnect principles within these topics that describe fundamental chemical phenomena. The six Big Ideas are as follows:

- Big Idea #1: The chemical elements are fundamental building materials of matter, and all matter can be understood in terms of arrangements of atoms. These atoms retrain their identity in chemical reactions.

- Big Idea #2: Chemical and physical properties of materials can be explained by the structure and the arrangement of atoms, ions, or molecules and the forces between them.

- Big Idea #3: Changes in matter involve the rearrangement and/or reorganization of atoms and/or the transfer of electrons.

- Big Idea #4: Rates of chemical reactions are determined by details of the molecular collisions.

- Big Idea #5: The laws of thermodynamics describe the essential role of energy and explain and predict the direction of changes in matter.

- Big Idea #6: Any bond or intermolecular attraction that can be formed can by broken. These two processes are in a dynamic competition, sensitive to initial conditions and external perturbations.

This book has been arranged to teach chemistry topics grouped to each of these Big Ideas.

HOW AP EXAMS ARE USED

Different colleges use AP Exam scores in different ways, so it is important that you go to a particular college's web site to determine how it uses AP Exam scores. The three items below represent the main ways in which AP Exam scores can be used:

- **College Credit.** Some colleges will give you college credit if you score well on an AP Exam. These credits count towards your graduation requirements, meaning that you can take fewer courses while in college. Given the cost of college, this could be quite a benefit, indeed.

- **Satisfy Requirements.** Some colleges will allow you to "place out" of certain requirements if you do well on an AP Exam, even if they do not give you actual college credits. For example, you might not need to take an introductory-level course, or perhaps you might not need to take a class in a certain discipline at all.

- **Admissions Plus.** Even if your AP Exam will not result in college credit or even allow you to place out of certain courses, most colleges will respect your decision to push yourself by taking an AP Course or even an AP Exam outside of a course. A high score on an AP Exam shows mastery of more difficult content than is taught in many high school courses, and colleges may take that into account during the admissions process.

OTHER RESOURCES

There are many resources available to help you improve your score on the Chemistry Exam, not the least of which are your **teachers.** If you are taking an AP class, you may be able to get extra attention from your teacher, such as obtaining feedback on your free response questions. If you are not in an AP course, reach out to a teacher who teaches Chemistry, and ask if the teacher will review your free response questions or otherwise help you with content.

Another wonderful resource is **AP Central,** the official site of the AP Exams. The scope of the information at this site is quite broad and includes:

- Course Description, which includes details on what content is covered and sample questions

- Sample questions for the new May 2014 exam

- Free response question prompts and multiple choice questions from previous years

The AP Central home page address is: **http://apcentral.collegeboard.com/apc/Controller.jpf.**

For up-to-date information about the ongoing changes to the AP Chemistry Exam Course, please visit: **http://advancesinap.collegeboard.org/science/chemistry.**

Finally, **The Princeton Review** offers tutoring and small group instruction. Our expert instructors can help you refine your strategic approach and add to your content knowledge. For more information, call 1-800-2REVIEW.

DESIGNING YOUR STUDY PLAN

As part of the Introduction, you identified some areas of potential improvement. Let's now delve further into your performance on Test 1, with the goal of developing a study plan appropriate to your needs and time commitment.

Read the answers and explanations associated with the Multiple Choice questions (starting at page 292). After you have done so, respond to the following questions:

- Review the Overview of Content Topics at page 8. Next to each topic, indicate your rank of the topic as follows: "1" means "I need a lot of work on this," "2" means "I need to beef up my knowledge," and "3" means "I know this topic well."

- How many days/weeks/months away is your exam?

- What time of day is your best, most focused study time?

- How much time per day/week/month will you devote to preparing for your exam?

- When will you do this preparation? (Be as specific as possible: Mondays & Wednesdays from 3 to 4 pm, for example)

- Based on the answers above, will you focus on strategy (Part Two) or content (Part Three) or both?

- What are your overall goals in using this book?

Part III
Test-Taking Strategies for the AP Chemistry Exam

PREVIEW ACTIVITY

Review the results of Test 1 and then respond to the following questions:

- How many multiple choice questions did you miss even though you knew the answers?

- On how many multiple choice did you guess blindly?

- How many multiple choice questions did you miss after eliminating some answers and guessing based on the remaining answers?

- Did you find any of the free-response questions easier or harder than the others—and, if so, why?

Chapter 1
How to
Approach
Multiple-Choice
Questions

THE BASICS

Section I of the test is composed of 60 multiple-choice questions, for which you are allotted 90 minutes. This part is worth 50 percent of your total score.

For this section, you will be given a periodic table of the elements and you may NOT use a calculator. The College Board says that this is because the new scientific calculators not only program and graph but also store information—and they are afraid you'll use this function to cheat!

On the multiple-choice section, you receive 1 point for a correct answer. There is no penalty for leaving a question blank or getting a question wrong.

PACING

According to the College Board, the multiple choice section of the AP Chemistry Exam covers more material than any individual student is expected to know. Nobody is expected to get a perfect or even near perfect score. What does that mean to you?

Use the Two-Pass System

Go through the multiple-choice section twice. The first time, do all the questions that you can get answers to immediately. That is, do the questions with little or no math and questions on chemistry topics in which you are well versed. Skip questions on topics that make you uncomfortable. Also, you want to skip the ones that look like number crunchers (even without a calculator, you may still be expected to crunch a few numbers). Circle the questions that you skip in your test booklet so you can find them easily during the second pass. Once you've done all the questions that come easily to you, go back and pick out the tough ones that you have the best shot at.

In general, the questions near the end of the section are tougher than the questions near the beginning. You should keep that in mind, but be aware that each person's experience will be different. If you can do acid-base questions in your sleep, but you'd rather have your teeth drilled than draw a Lewis diagram, you may find questions near the end of the section easier than questions near the beginning.

That's why the Two-Pass System is so handy. By using it, you make sure you get to see all the questions you can get right, instead of running out of time because you got bogged down on questions you couldn't do earlier in the test.

This brings us to another important point.

Don't Turn a Question into a Crusade!

Most people don't run out of time on standardized tests because they work too slowly. Instead, they run out of time because they spend half the test wrestling with two or three particular questions.

You should never spend more than a minute or two on any question. If a question doesn't involve calculation, then either you know the answer, you can make an educated guess, or you don't know the answer. Figure out where you stand on a question, make a decision, and move on.

Any question that requires more than two minutes worth of calculations probably isn't worth doing. Remember: Skipping a question early in the section is a good thing if it means that you'll have time to get two correct answers later on.

GUESSING

You get one point for every correct answer on the multiple-choice section. Guessing randomly neither helps you nor hurts you. Educated guessing, however, will help you.

Use Process of Elimination (POE) to Find Wrong Answers

There is a fundamental weakness to a multiple-choice test. The test makers must show you the right answer, along with three wrong answers. Sometimes seeing the right answer is all you need. Other times you may not know the right answer, but you may be able to identify one or two of the answers that are clearly wrong. Here is where you should use POE to take an educated guess.

Look at this hypothetical question.

1. Which of the following compounds will produce a purple solution when added to water?

   (A) Brobogdium rabelide
   (B) Diblythium perjuvenide
   (C) Sodium chloride
   (D) Carbon dioxide

You should have no idea what the correct answer is because three of these compounds are made up, but you do know something about the obviously wrong answers. You know that sodium chloride, choice (C), and carbon dioxide, choice (D), do not turn water purple. So, using POE, you have a 50 percent chance at guessing the correct answer. Now the odds are in your favor. Now you should guess.

Guess and Move On

Remember that you're guessing. Pondering the possible differences between bro-bogdium rabelide and diblythium perjuvenide is a waste of time. Once you've taken POE as far as it will go, pick your favorite letter and move on.

The multiple-choice section is the exact opposite of the free-response section. It's scored by a machine. There's no partial credit. The computer doesn't know, or care if you know, why an answer is correct. All the computer cares about is whether you blackened in the correct oval on your score sheet. You get the same number of points for picking (B) because you know (A) is wrong and (B) is a nicer letter than (C) or (D) as you would for picking (B) because you fully understood the subtleties of an electrochemical process.

ABOUT CALCULATORS

You will NOT be allowed to use a calculator on this section. That shouldn't worry you. All it means is that there won't be any questions in the section that you'll need a calculator to solve.

Most of the calculation problems will have fairly user-friendly numbers—that is, numbers with only a couple of significant digits, or things like "11.2 liters of gas at STP" or "160 grams of oxygen" or "a temperature increase from 27°C to 127°C." Sometimes these user-friendly numbers will actually point you toward the proper steps to take in your calculations.

Don't be afraid to make rough estimates as you do your calculations. Sometimes knowing that an answer is closer to 50 than to 500 will enable you to pick the correct answer on a multiple-choice test (if the answer choices are far enough apart). Once again, the rule against calculators works in your favor because the College Board will not expect you to do very precise calculations by hand.

There may be a couple of real number-crunching problems on the test. If you can recognize them quickly, these are good ones to skip. There's no point in spending five minutes crunching numbers to get one problem right if that time could be better used in getting three others correct later in the test.

REFLECT

Respond to the following questions:

- How long will you spend on multiple choice questions?

- How will you change your approach to multiple choice questions?

- What is your multiple choice guessing strategy?

Chapter 2
How to Approach
Free Response
Questions

CRACKING THE FREE RESPONSE SECTION

Section II is composed of a series of seven free response questions, all of which are required. You will be allotted 90 minutes to complete this section, which is worth 50 percent of your total score.

Cracking the Math Problems

You want to show the graders that you can do chemistry math, so here are some suggestions.

Show every step of your calculations on paper

This section is the opposite of multiple choice. You don't just get full credit for writing the correct answer. You get most of your points on this section for showing the process that got you to the answer. The graders give you partial credit when you show them that you know what you're doing. So even if you can do a calculation in your head, you should set it up and show it on the page.

By showing every step, or explaining what you're doing in words, you insure that you'll get all the partial credit possible, even if you screw up a calculation.

Include units in all your calculations

Scientists like units in calculations. Units make scientists feel secure. You'll get points for including them and you may lose points for leaving them out.

Remember significant figures

You can lose one point per question if your answer is off by more than one significant figure. Without getting too bent out of shape about it, try to remember that a calculation is only as accurate as the least accurate number in it.

The graders will follow your reasoning, even if you've made a mistake

Often, you are asked to use the result of a previous part of a problem in a later part. If you got the wrong answer in part (a) and used it in part (c), you can still get full credit for part (c), as long as your work is correct based on the number that you used. That's important, because it means that botching the first part of a question doesn't necessarily sink the whole question.

Remember the mean!

So let's say that you could complete only parts (a) and (b) on the required equilibrium problem. That's 4 or 5 points out of 10, tops. Are you doomed? Of course not. You're above average. If this test is hard on you, it's probably just as hard on everybody else. Remember: You don't need anywhere near a perfect score to get a 5, and you can leave half the test blank and still get a 4!

STRATEGIES FOR CRACKING THE FREE RESPONSE SECTION

This section is here to test whether you can translate chemistry into English. The term "essay" is a little misleading because all of these questions can be answered in two or three simple sentences, or with a simple diagram or two. Here are some tips for answering the seven essay questions on Part II.

Show that you understand the terms used in the question

If they ask you why sodium and potassium have differing first ionization energies, the first thing you should do is tell them what ionization energy is. That's probably worth the first point of partial credit. Then you should tell them how the differing structures of the atoms make for differing ionization energies. That leads to the next tip.

Take a step-by-step approach

Grading these tests is hard work. Breaking a question into parts in this way makes it easier on the grader, who must match your response to a set of guidelines he or she has been given that describe how to assign partial and full credit.

Each grader scores each test based on these rough guidelines that are established at the beginning of the grading period. For instance, if a grader has 3 points for the question about ionization energies, the points might be distributed the following way:

- One point for understanding ionization energy.
- One point for explaining the structural difference between sodium and potassium.
- One point for showing how this difference affects the ionization energy.

You can get all three points for this question if the grader thinks that all three concepts are addressed implicitly in your answer, but by taking a step by step approach, you improve your chances of explicitly addressing the things that a grader has been instructed to look for. Once again, grading these tests is hard work; graders won't know for sure if you understand something unless you tell them.

This leads us to an obvious point.

Write neatly

Even if writing neatly means working at half-speed. You can't get points for answers if the graders can't understand them. Of course, this applies to the rest of the free-response section as well.

The graders will follow your reasoning, even if you've made a mistake

Just like in the problems section, you might be asked to use the result of a previous part of a problem in a later part. If you decide (incorrectly) that an endothermic reaction in part (a) is exothermic, you can still get full credit in part (c) for your wrong answer about the reaction's spontaneity, as long as your answer in (c) is correct based on an exothermic reaction.

REFLECT

Respond to the following questions:

- How much time will you spend on the short free-response questions? What about the long free-response questions?

- What will you do before you begin writing your free response answers?

- Will you seek further help, outside of this book (such as a teacher, tutor, or AP Central), on how to approach the questions that you will see on the AP Chemistry exam?

Part IV
Content Review for the AP Chemistry Exam

Chapter 3
Big Idea #1: Atoms, Elements, and the Building Blocks of Matter

The chemical elements are fundamental building materials of matter, and all matter can be understood in terms of arrangements of atoms. These atoms retain their identity in chemical reactions.

THE PERIODIC TABLE

The most important tool you will use on this test is the Periodic Table of the Elements.

PERIODIC TABLE OF THE ELEMENTS

1 **H** 1.008																	**2** **He** 4.00
3 **Li** 6.94	**4** **Be** 9.01											**5** **B** 10.81	**6** **C** 12.01	**7** **N** 14.1	**8** **O** 16.00	**9** **F** 19.00	**10** **NE** 20.18
11 **Na** 22.99	**12** **Mg** 24.30											**13** **Al** 26.98	**14** **Si** 28.09	**15** **P** 30.97	**16** **S** 32.06	**17** **Cl** 35.45	**18** **Ar** 39.95
19 **K** 39.10	**20** **Ca** 40.08	**21** **Sc** 44.96	**22** **Ti** 47.90	**23** **V** 50.94	**24** **Cr** 52.00	**25** **Mn** 59.94	**26** **Fe** 55.85	**27** **Co** 58.93	**28** **Ni** 58.69	**29** **Cu** 63.55	**30** **Zn** 65.39	**31** **Ga** 69.72	**32** **Ge** 72.59	**33** **As** 74.92	**34** **Se** 78.96	**35** **Br** 79.90	**36** **Kr** 83.80
37 **Rb** 85.47	**38** **Sr** 87.62	**39** **Y** 88.91	**40** **Zr** 91.22	**41** **Nb** 92.91	**42** **Mo** 95.94	**43** **Tc** (98)	**44** **Ru** 101.1	**45** **Rh** 102.91	**46** **Pd** 106.42	**47** **Ag** 107.87	**48** **Cd** 112.41	**49** **In** 114.82	**50** **Sn** 118.71	**51** **Sb** 121.75	**52** **Te** 127.60	**53** **I** 126.91	**54** **Xe** 131.29
55 **Cs** 132.91	**56** **Ba** 137.33	**57** ***La** 138.91	**72** **Hf** 178.49	**73** **Ta** 180.95	**74** **W** 183.85	**75** **Re** 186.21	**76** **Os** 190.2	**77** **Ir** 192.2	**78** **Pt** 195.08	**79** **Au** 196.97	**80** **Hg** 200.59	**81** **Tl** 204.38	**82** **Pb** 207.2	**83** **Bi** 208.98	**84** **Po** (209)	**85** **At** (210)	**86** **Rn** (222)
87 **Fr** (223)	**88** **Ra** 226.02	**89** **†Ac** 227.03	**104** **Rf** (261)	**105** **Db** (262)	**106** **Sg** (266)	**107** **Bh** (264)	**108** **Hs** (277)	**109** **Mt** (268)	**110** **Ds** (271)	**111** **Rg** (272)							

*Lanthanide Series	**58** **Ce** 140.12	**59** **Pr** 104.91	**60** **Nd** 144.24	**61** **Pm** (145)	**62** **Sm** 150.4	**63** **Eu** 151.97	**64** **Gd** 157.25	**65** **Tb** 158.93	**66** **Dy** 162.50	**67** **Ho** 164.93	**68** **Er** 167.26	**69** **Tm** 169.93	**70** **Yb** 173.04	**71** **Lu** 174.97
†Actinide Series	**90** **Th** 232.04	**91** **Pa** 231.04	**92** **U** 238.03	**93** **X** (237)	**94** **Pu** (244)	**95** **Am** (243)	**96** **Cm** (247)	**97** **Bk** (247)	**98** **Cf** (251)	**99** **Es** (252)	**100** **Fm** (257)	**101** **Md** (258)	**102** **No** (259)	**103** **Lr** (262)

The periodic table gives you very basic but very important information about each element.

1. This is the **symbol** for the element; carbon, in this case. On the test, the symbol for an element is used interchangeably with the name of the element.
2. This is the **atomic number** of the element. The atomic number is the same as the number of protons in the nucleus of an element; it is also the same as the number of electrons surrounding the nucleus of an element in its neutral state.
3. This is the **molar mass** of the element. It's also called the **atomic weight**.

The identity of an atom is determined by the number of protons contained in its nucleus. The nucleus of an atom also contains neutrons. The mass number of an atom is the sum of its neutrons and protons. Atoms of an element with different numbers of neutrons are called isotopes; for instance, carbon-12, which contains 6 protons and 6 neutrons, and carbon-14, which contains 6 protons and 8 neutrons, are isotopes of carbon. The molar mass given on the periodic table is the average of the mass numbers of all known isotopes weighted by their percent abundance.

The molar mass of an element will give you a pretty good idea of the most common isotope of that element. For instance, the molar mass of carbon is 12.0 and about 99 percent of the carbon in existence is carbon-12.

The horizontal rows of the periodic table are called **periods**.

The vertical columns of the periodic table are called **groups**.

MOLES

The mole (**Avogadro's number**) is the most important number in chemistry, serving as a bridge that connects all the different quantities that you'll come across in chemical calculations. The coefficients in chemical reactions tell you about the reactants and products in terms of moles, so most of the stoichiometry questions you'll see on the test will be exercises in converting between moles and grams, liters, molarities, and other units.

Moles and Molecules

The definition of Avogadro's number gives you the information you need to convert between moles and individual molecules and atoms.

$$1 \text{ mole} = 6.022 \times 10^{23} \text{ molecules}$$

$$\text{Moles} = \frac{\text{molecules}}{\left(6.022 \times 10^{23}\right)}$$

Moles and Grams

Moles and grams can be related by using the atomic weights given in the periodic table. Atomic weights on the periodic table are given in terms of atomic mass units (amu), but an amu is the same as a gram per mole, so if 1 carbon atom weighs 12 amu, then 1 mole of carbon atoms weighs 12 grams.

You can use the relationship between amu and g/mol to convert between grams and moles by using the following equation:

$$\text{Moles} = \frac{\text{grams}}{\text{molar mass}}$$

Moles and Gases

We'll talk more about the ideal gas equation in Chapter 4, but for now, you should know that you can use it to calculate the number of moles of a gas if you know some of the gas's physical properties. All you need to remember at this point is that in the equation $PV = nRT$, n stands for moles of gas.

$$\text{Moles} = \frac{PV}{RT}$$

P = pressure (atm)
V = volume (L)
T = temperature (K)
R = the gas constant, 0.0821 L-atm/mol-K

The equation above gives the general rule for finding the number of moles of a gas. Many gas problems will take place at STP, or standard temperature and pressure, where P = 1 atmosphere and T = 273 K. At STP, the situation is much simpler and you can convert directly between the volume of a gas and the number of moles. That's because at STP, one mole of gas always occupies 22.4 liters.

$$\text{Moles} = \frac{\text{liters}}{\left(22.4\,\text{L/mol}\right)}$$

Moles and Solutions

We'll talk more about molarity and molality in Chapter 4, but for now you should realize that you can use the equations that define these common measures of concentration to find the number of moles of solute in a solution. Just rearrange the equations to isolate moles of solute.

Moles = (molarity)(liters of solution)
Moles = (molality)(kilograms of solvent)

Percent Composition

To solve many problems on the exam, you will need to use percent composition, or mass percents. Percent composition is the percent by mass of each element that makes up a compound. It is calculated by dividing the mass of each element or component in a compound by the total molar mass for the substance.

Empirical and Molecular Formulas

You will also need to know how to determine the empirical and molecular formula of a compound given masses or mass percents of the components of that compound. Remember that the empirical formula represents the simplest ratio of one element to another in a compound (e.g., CH_2O), while the molecular formula represents the actual formula for the substance (e.g., $C_6H_{12}O_6$).

ELECTRONS

Electronic Configurations and the Periodic Table

The positions of the electrons in relation to the nucleus are described by their **quantum numbers**. Each electron has four quantum numbers that apply to its **shell**, **subshell**, **orbital**, and **spin**.

Shells: n = 1, 2, 3...

In a hydrogen atom, the principal quantum number, or shell, of an electron determines its average distance from the nucleus as well as its energy. So electrons in shells with higher values are farther away on average from the nucleus and will have more energy and less stability than electrons in shells with lower values.

Subshells: l = 0, 1, 2...

The angular momentum quantum number, or subshell, describes the shape of an electron's orbital.

- The first shell (n = 1) has one subshell: s, or l = 0.
- The second shell (n = 2) has two subshells: s (l = 0), and p (l = 1).
- The third shell (n = 3) has three subshells: s (l = 0), p (l = 1), and d (l = 2).

The orbitals of s subshells are spherical, while the orbitals of p subshells are dumbbell shaped.

Orbitals: $m_l = ...-1, 0, +1...$

The magnetic quantum number, or orbital, describes the orientation of the orbital in space. Roughly, that means it describes whether the path of the electron lies mostly on the *x*, *y*, or *z* axis of a three-dimensional grid.

- The *s* subshell ($l = 0$) has one orbital: $m_l = 0$.
- The *p* subshell ($l = 1$) has three orbitals: $m_l = -1$, $m_l = 0$, and $m_l = +1$.
- The *d* subshell ($l = 2$) has five orbitals: $m_l = -2, -1, 0, 1,$ and 2.

Spin: $m_s = +\dfrac{1}{2}, -\dfrac{1}{2}$

Each orbital can contain two electrons: one with a positive spin and one with a negative spin.

Here are a couple of graphical ways of looking at quantum numbers.

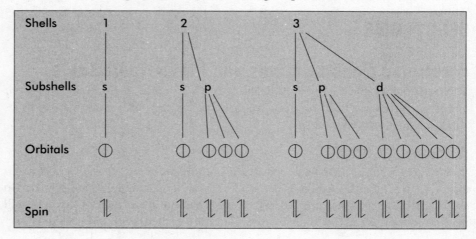

	The Quantum Numbers for levels 1–4			The number of electrons in the states	
n	*l*	*m*		In the subshell	In the level
1	0 (s)	0		2	2
2	0 (s)	0		2	8
	1 (p)	–1, 0, +1		6	
3	0 (s)	0		2	18
	1 (p)	–1, 0, +1		6	
	2 (d)	–2, –1, 0, +1, +2		10	
4	0 (s)	0		2	32
	1 (p)	–1, 0, +1		6	
	2 (d)	–2, –1, 0, +1, +2		10	
	3 (f)	–3, –2, –1, 0, +1, +2, +3		14	

The Aufbau Principle

The **Aufbau principle** states that when building up the electron configuration of an atom, electrons are placed in orbitals, subshells, and shells in order of increasing energy.

The Pauli Exclusion Principle

The **Pauli exclusion principle** states that within an atom, no two electrons can have the same set of quantum numbers. So, each electron in any atom has its own distinct set of four quantum numbers.

Hund's Rule

Hund's rule says that when an electron is added to a subshell, it will always occupy an empty orbital if one is available. Electrons always occupy orbitals singly if possible and pair up only if no empty orbitals are available.

Watch how the $2p$ subshell fills as we go from boron to neon.

	1s	2s	2p
Boron	⇅	⇅	↑
Carbon	⇅	⇅	↑ ↑
Nitrogen	⇅	⇅	↑ ↑ ↑
Oxygen	⇅	⇅	⇅ ↑ ↑
Fluorine	⇅	⇅	⇅ ⇅ ↑
Neon	⇅	⇅	⇅ ⇅ ⇅

Electrons and Energy

The positively-charged nucleus is always pulling at the negatively-charged electrons around it, and the electrons have potential energy that increases with their distance from the nucleus. It works the same way that the gravitational potential energy of a brick on the third floor of a building is greater than the gravitational potential energy of a brick nearer to ground level.

The energy of electrons, however, is **quantized.** That's important. It means that electrons can exist only at specific energy levels, separated by specific intervals. It's kind of like if the brick in the building could be placed only on the first, second, or third floor of the building, but not in-between.

The quantized energy of an electron in a hydrogen atom can be found if you know its principal quantum number or shell.

Energy of an Electron

$$E_n = \frac{-2.178 \times 10^{-18}}{n^2} \text{ joules}$$

E_n = the energy of the electron

n = the principal quantum number of the electron

When atoms absorb energy in the form of **electromagnetic radiation**, electrons jump to higher energy levels. When electrons drop from higher to lower energy levels, atoms give off energy in the form of electromagnetic radiation.

The relationship between the change in energy level of an electron and the electromagnetic radiation absorbed or emitted is given below.

Energy and Electromagnetic Radiation

$$\Delta E = h v = \frac{hc}{\lambda}$$

ΔE = energy change

h = Planck's constant, 6.63×10^{-34} joule-sec

v = frequency of the radiation

λ = wavelength of the radiation

c = the speed of light, 3.00×10^8 m/sec ($c = \lambda f$)

The energy level changes for the electrons of a particular atom are always the same, so atoms can be identified by their emission and absorption spectra.

Names and Theories

Dalton's Elements

In the early 1800s, John Dalton presented some basic ideas about atoms that we still use today. He was the first to say that there are many different kinds of atoms, which he called elements. He said that these elements combine to form compounds and that these compounds always contain the same ratios of elements. Water (H_2O), for instance, always has two hydrogen atoms for every oxygen atom. He also said that atoms are never created or destroyed in chemical reactions.

Development of the Periodic Table

In 1869, Dmitri Mendeleev and Lothar Meyer independently proposed arranging the elements into early versions of the periodic table, based on the trends of the known elements.

Thomson's Experiment

In the late 1800s, J. J. Thomson watched the deflection of charges in a cathode ray tube and put forth the idea that atoms are composed of positive and negative charges. The negative charges were called electrons, and Thomson guessed that they were sprinkled throughout the positively charged atom like chocolate chips sprinkled throughout a blob of cookie dough.

Millikan's Experiment

Robert Millikan was able to calculate the charge on an electron by examining the behavior of charged oil drops in an electric field.

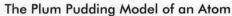

The Plum Pudding Model of an Atom

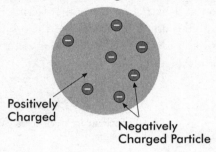

Positively Charged

Negatively Charged Particle

Rutherford's Experiment

In the early 1900s, Ernest Rutherford fired alpha particles at gold foil and observed how they were scattered. This experiment led him to conclude that all of the positive charge in an atom was concentrated in the center and that an atom is mostly empty space. This led to the idea that an atom has a positively charged nucleus, which contains most of the atom's mass, and that the tiny, negatively charged electrons travel around this nucleus.

Quantum Theory

Max Planck figured out that electromagnetic energy is quantized. That is, for a given frequency of radiation (or light), all possible energies are multiples of a certain unit of energy, called a quantum (mathematically, that's $E = h\nu$). So, energy changes do not occur smoothly but rather in small but specific steps.

The Bohr Model

Neils Bohr took the quantum theory and used it to predict that electrons orbit the nucleus at specific, fixed radii, like planets orbiting the Sun. The Bohr model worked for atoms and ions with one electron but not for more complex atoms.

Atom absorbs energy

Atom emits energy

The Heisenberg Uncertainty Principle

Werner Heisenberg said that it is impossible to know both the position and momentum of an electron at a particular instant. In terms of atomic structure, this means that electron orbitals do not represent specific orbits like those of planets. Instead, an electron orbital is a probability function describing the possibility that an electron will be found in a region of space.

Probability Density for Electron in Hydrogen Atom

The de Broglie Hypothesis

Louis de Broglie said that all matter has wave characteristics. This is important because sometimes the behavior of electrons is better described in terms of waves than particles.

There is a simple relationship between an electron's wave and particle characteristics.

The de Broglie Equation

$$\lambda = \frac{h}{mv}$$

λ = wavelength associated with a particle

m = mass of the particle

v = speed of the particle

mv = p = momentum of the particle

h = Planck's constant, 6.63×10^{-34} joule-sec

De Broglie's hypothesis is useful for very small particles, such as electrons. For larger particles, the wavelength becomes too small to be of interest.

PERIODIC TRENDS

You can make predictions about certain behavior patterns of an atom and its electrons based on the position of the atom in the periodic table. All the periodic trends can be understood in terms of three basic rules.

1. Electrons are attracted to the protons in the nucleus of an atom.
   a. The closer an electron is to the nucleus, the more strongly it is attracted.
   b. The more protons in a nucleus, the more strongly an electron is attracted.
2. Electrons are repelled by other electrons in an atom. So, if other electrons are between a valence electron and the nucleus, the valence electron will be less attracted to the nucleus. That's called shielding.
3. Completed shells (and to a lesser extent, completed subshells) are very stable. Atoms prefer to add or subtract valence electrons to create complete shells if possible.

The atoms in the left-hand side of the periodic table are called metals. Metals give up electrons when forming bonds. Most of the elements in the table are metals. The elements in the upper right-hand portion of the table are called nonmetals. Nonmetals generally gain electrons when forming bonds. The metallic character of the elements decreases as you move from left to right across the periodic table. The elements in the borderline between metal and nonmetal, such as silicon and arsenic, are called metalloids.

Atomic Radius

The atomic radius is the approximate distance from the nucleus of an atom to its valence electrons.

Moving from Left to Right Across a Period (Li to Ne, for Instance), Atomic Radius Decreases

Moving from left to right across a period, protons are added to the nucleus, so the valence electrons are more strongly attracted to the nucleus; this decreases the atomic radius. Electrons are also being added, but they are all in the same shell at about the same distance from the nucleus, so there is not much of a shielding effect.

Moving Down a Group (Li to Cs, for Instance), Atomic Radius Increases

Moving down a group, shells of electrons are added to the nucleus. Each shell shields the more distant shells from the nucleus and the valence electrons get farther away from the nucleus. Protons are also being added, but the shielding effect of the negatively charged electron shells cancels out the added positive charge.

Cations (Positively Charged Ions) Are Smaller than Atoms

Generally, when electrons are removed from an atom to form a cation, the outer shell is lost, making the cation smaller than the atom. Also, when electrons are removed, electron-electron repulsions are reduced, allowing all of the remaining valence electrons to move closer to the nucleus.

Anions (Negatively Charged Ions) Are Larger than Atoms

When an electron is added to an atom, forming an anion, electron-electron repulsions increase, causing the valence electrons to move farther apart, which increases the radius.

Ionization Energy

Electrons are attracted to the nucleus of an atom, so it takes energy to remove an electron. The energy required to remove an electron from an atom is called the first ionization energy. Once an electron has been removed, the atom becomes a positively charged ion. The energy required to remove the next electron from the ion is called the second ionization energy, and so on.

Moving from Left to Right Across a Period, Ionization Energy Increases

Moving from left to right across a period, protons are added to the nucleus, which increases its positive charge. For this reason, the negatively charged valence electrons are more strongly attracted to the nucleus, which increases the energy required to remove them. Electrons are also being added, and the shielding effect provided by the filling of the s subshell causes a slight deviation in the trend in moving from Group 2A to Group 3A. There is also a slight deviation when the electrons in the p subshell start to pair up, so oxygen has a slightly lower first ionization energy than nitrogen does.

Moving Down a Group, Ionization Energy Decreases

Moving down a group, shells of electrons are added to the nucleus. Each inner shell shields the more distant shells from the nucleus, reducing the pull of the nucleus on the valence electrons and making them easier to remove. Protons are also being added, but the shielding effect of the negatively charged electron shells cancels out the added positive charge.

The Second Ionization Energy Is Greater than the First Ionization Energy

When an electron has been removed from an atom, electron-electron repulsion decreases and the remaining valence electrons move closer to the nucleus. This increases the attractive force between the electrons and the nucleus, increasing the ionization energy.

As Electrons Are Removed, Ionization Energy Increases Gradually Until a Shell Is Empty, Then Makes a Big Jump

- For each element, when the valence shell is empty, the next electron must come from a shell that is much closer to the nucleus, making the ionization energy for that electron much larger than for the previous ones.
- For Na, the second ionization energy is much larger than the first.
- For Mg, the first and second ionization energies are comparable, but the third is much larger than the second.
- For Al, the first three ionization energies are comparable, but the fourth is much larger than the third.

Electron Affinity

Electron affinity is a measure of the change in energy of an atom when an electron is added to it. When the addition of an electron makes the atom more stable, energy is given off. This is true for most of the elements. When the addition of an electron makes the atom less stable, energy must be put in; that's because the added electron must be placed in a higher energy level, making the element less stable. This is the case for elements with full subshells, like the alkaline earths and the noble gases.

When moving from left to right across a period, the energy given off when an electron is added increases. Electron affinities don't change very much moving down a group.

Electronegativity

Electronegativity refers to how strongly the nucleus of an atom attracts the electrons of other atoms in a bond. Electronegativities of elements are estimated based on ionization energies and electron affinities, and they basically follow the same trends.

- Moving from left to right across a period, electronegativity increases.
- Moving down a group, electronegativity decreases.

The various periodic trends, which don't include the noble gases, are summarized in the diagram below.

NUMBER OF PROTONS INCREASES
Atomic radius decreases
Ionization energy increases
Electronegativity increases

SHIELDING INCREASES
Atomic radius increases
Ionization energy decreases
Electronegativity decreases

CHAPTER 3 QUESTIONS

Multiple-Choice Questions

Questions 1–4

    (A) C
    (B) N
    (C) F
    (D) Ne

1. This is the most electronegative element.

2. The nuclear decay of an isotope of this element is used to measure the age of archaeological artifacts.

3. All of the electrons in this element are spin-paired.

4. This element, present as a diatomic gas, makes up most of the Earth's atmosphere.

Questions 5–7

    (A) Hg
    (B) Si
    (C) Cu
    (D) Zn

5. This element is commonly used in the manufacture of semiconductors.

6. This element is a liquid at room temperature.

7. After oxygen, this is by far the most common element in the Earth's crust.

Questions 8-10

    (A) Moles
    (B) Liters
    (C) Grams
    (D) Atmospheres

8. One mole of solid zinc has a mass of 65.39 of these.

9. These units can be calculated by dividing a quantity by 6.02×10^{23}.

10. Four grams of helium gas occupy 22.4 of these at standard temperature and pressure.

11. What is the most likely electron configuration for a sodium ion in its ground state?

(A) $1s^2\,2s^2\,2p^5$
(B) $1s^2\,2s^2\,2p^6$
(C) $1s^2\,2s^2\,2p^6\,3s^1$
(D) $1s^2\,2s^2\,2p^5\,3s^2$

12. Which of the following statements is true regarding sodium and chlorine?

(A) Sodium has greater electronegativity and a larger first ionization energy.
(B) Sodium has a larger first ionization energy and a larger atomic radius.
(C) Chlorine has a larger atomic radius and a greater electronegativity.
(D) Chlorine has greater electronegativity and a larger first ionization energy.

13. Which of the following rules states that no two electrons in an atom can have the same set of quantum numbers?

(A) Hund's rule
(B) The Heisenberg uncertainty principle
(C) The Pauli exclusion principle
(D) The de Broglie hypothesis

14. Which of the following is true of the alkali metal elements?

(A) They usually take the +2 oxidation state.
(B) They have oxides that act as acid anhydrides.
(C) They form covalent bonds with oxygen.
(D) They are generally found in nature in compounds.

15. Which of the following nuclei has 3 more neutrons than protons? (Remember: The number before the symbol indicates atomic mass.)

(A) ^{11}B
(B) ^{37}Cl
(C) ^{24}Mg
(D) ^{70}Ga

16. Which of the following ions has the smallest ionic radius?

(A) O^{2-}
(B) F^-
(C) Mg^{2+}
(D) Al^{3+}

17. Which of the following represents the energy of the single electron in a hydrogen atom when it is in the $n = 4$ state?

(A) $\left(\dfrac{-2.178\ \times\ 10^{-18}}{2}\right)$ joules

(B) $\left(\dfrac{-2.178\ \times\ 10^{-18}}{4}\right)$ joules

(C) $\left(\dfrac{-2.178\ \times\ 10^{-18}}{8}\right)$ joules

(D) $\left(\dfrac{-2.178\ \times\ 10^{-18}}{16}\right)$ joules

18. When an electron in a hydrogen atom makes the transition from the $n = 4$ state to the $n = 2$ state, blue light with a wavelength of 434 nm is emitted. Which of the following expressions gives the energy released by the transition?

(A) $\dfrac{\left(6.63 \times 10^{-34}\right)\left(3.00 \times 10^{8}\right)}{\left(4.34 \times 10^{-7}\right)}$ joules

(B) $\dfrac{\left(6.63 \times 10^{-34}\right)\left(4.34 \times 10^{-7}\right)}{\left(3.00 \times 10^{8}\right)}$ joules

(C) $\dfrac{\left(6.63 \times 10^{-34}\right)}{\left(4.34 \times 10^{-7}\right)\left(3.00 \times 10^{8}\right)}$ joules

(D) $\dfrac{\left(4.34 \times 10^{-7}\right)}{\left(6.63 \times 10^{-34}\right)\left(3.00 \times 10^{8}\right)}$ joules

19. The ionization energies for an element are listed in the table below.

First	Second	Third	Fourth	Fifth
8 eV	15 eV	80 eV	109 eV	141 eV

Based on the ionization energy table, the element is most likely to be

(A) sodium.
(B) magnesium.
(C) aluminum.
(D) silicon.

20. A researcher listed the first five ionization energies for a silicon atom in order from first to fifth. Which of the following lists corresponds to the ionization energies for silicon?

(A) 780 kJ, 13,675 kJ, 14,110 kJ, 15,650 kJ, 16,100 kJ
(B) 780 kJ, 1,575 kJ, 14,110 kJ, 15,650 kJ, 16,100 kJ
(C) 780 kJ, 1,575 kJ, 3,220 kJ, 15,650 kJ, 16,100 kJ
(D) 780 kJ, 1,575 kJ, 3,220 kJ, 4,350 kJ, 16,100 kJ

FREE RESPONSE QUESTIONS

1. Explain each of the following in terms of atomic and molecular structures and/or forces.

   (a) The first ionization energy for magnesium is greater than the first ionization energy for calcium.

   (b) The first and second ionization energies for calcium are comparable, but the third ionization energy is much greater.

   (c) Solid sodium conducts electricity, but solid sodium chloride does not.

   (d) The first ionization energy for aluminum is lower than the first ionization energy for magnesium.

2. Use your knowledge of the periodic table of the elements to answer the following questions.

   (a) Explain the trend in electronegativity from P to S to Cl.

   (b) Explain the trend in electronegativity from Cl to Br to I.

   (c) Explain the trend in atomic radius from Li to Na to K.

   (d) Explain the trend in atomic radius from Al to Mg to Na.

3 Use your knowledge of atomic theory to answer the following questions.

   (a) State the Heisenberg uncertainty principle.

   (b) The absorption spectrum of a hydrogen atom contains dark bands at specific wavelengths. The emission spectrum of a hydrogen atom contains bright bands at the same wavelengths. Explain what causes these bright and dark bands at specific wavelengths.

   (c) Explain why the addition of an electron to a chlorine atom is an exothermic process and the addition of an electron to a magnesium atom is an endothermic process.

   (d) Explain how the valence electron configuration of sulfur is consistent with the existence of Na_2S and SF_6.

CHAPTER 3 ANSWERS AND EXPLANATIONS

Multiple-Choice Questions

1. **C** Fluorine, which needs one electron to complete its outer shell, is the most electronegative element, with an electronegativity of 4.0.

2. **A** In carbon dating, the ratio of carbon-14 to carbon-12 in an organic artifact is used to determine the age of the artifact.

3. **D** Neon's second shell is complete, so all of its electrons are spin-paired.

4. **B** Nitrogen gas (N_2) makes up 78 percent of the Earth's atmosphere. Oxygen is next at 21 percent.

5. **B** Silicon (Si) is used in semiconductor technology because it has properties that lie in-between metals and nonmetals.

6. **A** Mercury (Hg) is unusual among metals in that it is a liquid at room temperature. Its melting point is −39 degrees Celsius.

7. **B** Silicon makes up 26 percent of the Earth's crust by weight (oxygen makes up 50 percent). In fact, compounds including silicon and oxygen make up nearly all rocks and soils.

8. **C** The units for atomic weight are grams/mole.

9. **A** A mole is equal to 6.02×10^{23}.

10. **B** Four grams of helium is a mole. A mole of gas occupies 22.4 liters at STP.

11. **B** Neutral sodium in its ground state has the electron configuration shown in choice (C). Sodium forms a bond by giving up its one valence electron and becoming a positively charged ion with the same electron configuration as neon.

12. **D** As we move from left to right across the periodic table within a single period (from sodium to chlorine), we add protons to the nuclei, which progressively increases the pull of each nucleus on its electrons. So chlorine will have a higher first ionization energy, greater electronegativity, and a smaller atomic radius.

13. **C** The Pauli exclusion principle states that no two electrons in an atom can have the same set of quantum numbers.

About the other answers:

(A) Hund's rule states that within a subshell, electrons will be placed in empty orbitals while they are available and will start to pair up in orbitals only when no more empty orbitals are available.

(B) The Heisenberg uncertainty principle states that it is impossible to know with certainty both the position and momentum of a particle at the same moment.

(D) The de Broglie hypothesis relates the wave and particle properties of matter, using the following equation:

$$\text{Wavelength} = \frac{h}{\text{momentum}}$$

h is Planck's constant, 6.63×10^{-34} joule-sec

14. **D** The alkali metals (Li, Na, K...) are extremely reactive and are found in nature almost exclusively in compounds.

As for the other answers:

Choice (A) is wrong because the alkali metals take the +1 oxidation state. Choice (B) is wrong because alkali metal oxides are basic anhydrides; that is, they form basic solutions in water. Choice (C) is wrong because they form ionic bonds with oxygen. Choice (E) is wrong because they have only one valence electron, so they have relatively small first ionization energies.

15. **B** The atomic mass is the sum of the neutrons and protons in any atom's nucleus. Since atomic number, which indicates the number of protons, is unique to each element we can subtract this from the weight to find the number of neutrons. The atomic number of B is 5; hence, the ^{11}B has 6 neutrons, only 1 in excess. The atomic number of Cl is 17, so ^{37}Cl has 20 neutrons, 3 in excess. The atomic number of Mg is 12, so ^{24}Mg has the same number of neutrons as protons. The element Ga has an atomic number of 31, meaning there are 39 neutrons in the given nucleus and the atomic number for F is 9, meaning 10 neutrons in ^{19}F.

16. **D** All of the ions listed have the same electron configuration as neutral neon. Al^{3+} has the most protons, so its electrons will experience greater attractive force from the nucleus, resulting in the smallest ions.

17. **D** When you know the principal quantum number, you must use the formula for the energy of an electron, which is shown below.

$$E_n = \left(\frac{-2.178 \times 10^{-18}}{n^2} \right)$$

The variable n stands for the principal quantum number; in this case, n = 4.

18. **A** When you know the wavelength emitted, you must use the formula for the energy of an electron transition, which is shown below. Remember: A nanometer is equal to 10^{-9} meters.

$$E_n = \frac{hc}{\lambda} = \frac{\left(6.63 \times 10^{-34} \right) \left(3.00 \times 10^8 \right)}{\left(4.34 \times 10^{-7} \right)}$$

19. **B** The ionization energy will show a large jump when enough electrons have been removed to leave a stable shell. In this case, the jump occurs between the second and third electrons removed, so the element is stable after two electrons are removed. Magnesium (Mg) is the only element on the list with exactly two valence electrons.

20. **D** The ionization energy will show a large jump when enough electrons have been removed to leave a stable shell. Silicon has four valence electrons, so we would expect to see a large jump in ionization energy after the fourth electron has been removed. That's choice (D).

Free Response Questions

1. (a) Ionization energy is the energy required to remove an electron from an atom. The outermost electron in Ca is at the 4s energy level. The outermost electron in Mg is at the 3s level. The outermost electron in Ca is at a higher energy level and is more shielded from the nucleus, making it easier to remove.

   (b) Calcium has two electrons in its outer shell. The second ionization energy will be larger than the first but still comparable because both electrons are being removed from the same energy level. The third electron is much more difficult to remove because it is being removed from a lower energy level, so it will have a much higher ionization energy than the other two.

(c) Solid sodium exhibits metallic bonding, in which the positively charged sodium ions are held together by a sea of mobile, delocalized electrons. These electrons move freely from nucleus to nucleus, making solid sodium a good conductor.

Sodium chloride exhibits ionic bonding, in which positively charged sodium ions and negatively charged chlorine ions hold fixed places in a crystal lattice. The electrons are localized around particular nuclei and are not free to move about the lattice. This makes solid sodium chloride a bad conductor of electricity.

(d) The valence electron to be removed from magnesium is located in the completed 3s subshell, while the electron to be removed from aluminum is the lone electron in the 3p subshell. It is easier to remove the electron from the higher-energy 3p subshell than from the lower energy (completed) 3s subshell, so the first ionization energy is lower for aluminum.

2. (a) Electronegativity is the pull of the nucleus of one atom on the electrons of other atoms; it increases from P to S to Cl because nuclear charge increases. This is because as you from move left to right across the periodic table, atomic radii decrease in size. Increasing nuclear charge means that Cl has the most positively charged nucleus of the three and will exert the greatest pull on the electrons of other atoms.

(b) Electronegativity is the pull of the nucleus of one atom on the electrons of other atoms; it decreases from Cl to Br to I because electron shells are added. The added electron shells shield the nucleus, causing it to have less of an effect on the electrons of other atoms. Therefore, iodine will exert the least pull on the electrons of other atoms.

(c) Atomic radius increases from Li to Na to K because electrons are being added in higher energy levels, which are farther away from the nucleus; therefore, the K atom is the largest of the three.

(d) Atomic radius increases from Al to Mg to Na because protons are being removed from the nucleus while the energy levels of the valence electrons remains unchanged. If there are fewer positive charges in the nucleus, the electrons of Na will be less attracted to the nucleus and will remain farther away.

3. (a) A straightforward statement of the Heisenberg uncertainty principle is as follows: It is impossible to know with certainty both the position and momentum of a particle at one moment.

    (b) When a hydrogen atom absorbs energy, its electrons jump to higher energy levels. The absorbed energy shows up as a dark area on the absorption spectrum.

    A hydrogen atom gives off energy when its electrons jump back down to lower energy levels. Electromagnetic waves, which are emitted at these jumps, show up as bright areas on the emission spectrum.

    In an atom, energy is quantized, which means that electrons can exist only at specific energy levels. When an electron jumps from one energy level to another, it will always emit or absorb exactly the same amount of energy, and because $\Delta E = h\nu$, for a particular energy change, radiation of the same frequency will always be emitted or absorbed.

    (c) When an electron is added to chlorine, the chlorine ion created has a complete valence shell, which is an extremely stable, low energy configuration. When something becomes more stable, energy is given off, making the process exothermic.

    When an electron is added to a magnesium atom, it must be placed by itself in the $3p$ energy level. Adding an electron in a higher energy level makes the magnesium ion created more energetic and less stable, which means that the process is endothermic.

    (d) Sulfur has six valence electrons in its outer shell. In Na_2S, sulfur gains two electrons to give its outer shell a complete octet. In SF_6, sulfur uses sp^3d^2 hybridization to share all six of its valence electrons with fluorine atoms.

Chapter 4
Big Idea #2:
Bonding and Phases

Chemical and physical properties of materials can be explained by the structure and the arrangement of atoms, ions, or molecules and the forces between them.

COULOMB'S LAW

All bonds occur because of electrostatic attractions. Atoms stick together to form molecules, and atoms and molecules stick together to form liquids or solids because the negatively charged electrons of one atom are attracted to the positively charged nucleus of another atom.

Electrostatic forces are governed by Coulomb's law, and the entire study of bonding comes down to understanding how Coulomb's law applies to different chemical situations.

Let's take a look at Coulomb's law.

Coulomb's Law

Attractive force is proportional to $\dfrac{(+q)(-q)}{r^2}$

$+q$ = magnitude of the positive charge
$-q$ = magnitude of the negative charge
r = distance between the charges

You should be able to extrapolate two things from the formula above.

- Bigger charges mean stronger bonds; smaller charges mean weaker bonds.
- Charges close together mean stronger bonds; charges far apart mean weaker bonds.

BONDS OVERVIEW

Atoms join to form molecules because atoms like to have a full outer shell of electrons. This usually means having eight electrons in the outer shell. So atoms with too many or too few electrons in their valence shells will find one another and pass the electrons around until all the atoms in the molecule have stable outer shells. Sometimes an atom will give up electrons completely to another atom, forming an ionic bond. Sometimes atoms share electrons, forming covalent bonds.

Ionic Bonds

An ionic solid is held together by the electrostatic attractions between ions that are next to one another in a lattice structure. They often occur between metals and nonmetals. In an ionic bond, electrons are not shared. Instead, one atom gives up electrons and becomes a positively charged ion while the other atom accepts electrons and becomes a negatively charged ion.

The two ions in an ionic bond are held together by electrostatic forces. In the diagram below, a sodium atom has given up its single valence electron to a chlorine atom, which has seven valence electrons and uses the electron to complete its outer shell (with eight). The two atoms are then held together by the positive and negative charges on the ions.

$$\left[\text{Na} \right]^+ \left[:\ddot{\text{C}}\text{l}: \right]^-$$

The same electrostatic attractions that hold together the ions in a molecule of NaCl hold together a crystal of NaCl, so there is no real distinction between the molecules and the solid. Ionic bonds are strong, and substances held together by ionic bonds have high melting and boiling points.

Coulomb's law states that more highly charged ions will form stronger bonds than less highly charged ions and smaller ions will form stronger bonds than larger ions. So an ionic bond composed of ions with +2 and −2 charges will be stronger than a bond composed of ions with +1 and −1 charges. Also from Coulomb's law, we know that the smaller the ions in an ionic bond, the stronger the bond. This is because a small ionic radius allows the charges to get closer together and increases the force between them.

In an ionic solid, each electron is localized around a particular atom, so electrons do not move around the lattice; this makes ionic solids poor conductors of electricity. Ionic liquids, however, do conduct electricity because the ions themselves are free to move about in the liquid phase, although the electrons are still localized around particular atoms. Salts are held together by ionic bonds.

Covalent Bonds

In a covalent bond, two atoms share electrons. Each atom counts the shared electrons as part of its valence shell. In this way, both atoms achieve complete outer shells.

In the diagram below, two fluorine atoms, each of which has seven valence electrons and needs one electron to complete its valence shell, form a covalent bond. Each atom donates an electron to the bond, which is considered to be part of the valence shell of both atoms.

$$:\ddot{\text{F}}\cdot \ + \ \cdot\ddot{\text{F}}: \ \Rightarrow \ :\ddot{\text{F}}:\ddot{\text{F}}:$$

The number of covalent bonds an atom can form is the same as the number of electrons in its valence shell.

The first covalent bond formed between two atoms is called a sigma (σ) bond. All single bonds are sigma bonds. If additional bonds between the two atoms are formed, they are called pi (π) bonds. The second bond in a double bond is a pi bond and the second and third bonds in a triple bond are also pi bonds. Double and triple bonds are stronger and shorter than single bonds, but they are not twice or triple the strength.

Summary of Multiple Bonds			
Bond type:	Single	Double	Triple
Bond designation:	One sigma (s)	One sigma (σ) and one pi (π)	One sigma (σ) and two pi (π)
Bond order:	One	Two	Three
Bond length:	Longest	Intermediate	Shortest
Bond energy:	Least	Intermediate	Greatest

Single bonds have one sigma (σ) bond and a bond order of one. The single bond has the longest bond length and the least bond energy.

Polarity

In the F_2 molecule shown above, the two fluorine atoms share the electrons equally, but that's not usually the case in molecules. Usually, one of the atoms (the more electronegative one) will exert a stronger pull on the electrons in the bond—not enough to make the bond ionic, but enough to keep the electrons on one side of the molecule more than on the other side. This gives the molecule a dipole. That is, the side of the molecule where the electrons spend more time will be negative and the side of the molecule where the electrons spend less time will be positive.

Dipole Moment

The polarity of a molecule is measured by the dipole moment. The larger the dipole moment, the more polar the molecule. The greater the charge at the ends of the dipole and the greater the distance between the charges, the greater the value of the dipole moment.

Intermolecular Forces

Sometimes the bonds that hold the atoms or ions in liquids and solids together are the same strong bonds that hold the atoms or ions together in molecules. Intermolecular forces exist only with covalently bonded molecules. We will also discuss metallic and network forces here.

Network (Covalent) Bonds

In a network solid, atoms are held together in a lattice of covalent bonds. You can visualize a network solid as one big molecule. Network solids are very hard and have very high melting and boiling points.

The electrons in a network solid are localized in covalent bonds between particular atoms, so they are not free to move about the lattice. This makes network solids poor conductors of electricity.

The most commonly seen network solids are compounds of carbon (diamond) and silicon (SiO_2—quartz). The hardness of diamond is due to the tetrahedral network structure formed by carbon atoms whose electrons are configured in sp^3 hybridization. The three-dimensional complexity of the tetrahedral network means that there are no natural seams along which a diamond can be broken.

Hydrogen Bonds

Hydrogen bonds are similar to dipole–dipole attractions. In a hydrogen bond, the positively charged hydrogen end of a molecule is attracted to the negatively charged end of another molecule containing an extremely electronegative element (fluorine, oxygen, or nitrogen—F, O, N).

Hydrogen bonds are much stronger than dipole–dipole forces because when a hydrogen atom gives up its lone electron to a bond, its positively charged nucleus is left virtually unshielded. Substances that have hydrogen bonds, like water and ammonia, have higher melting and boiling points than substances that are held together by dipole–dipole forces.

Water is less dense as a solid than as a liquid because its hydrogen bonds force the molecules in ice to form a crystal structure, which keeps them farther apart than they are in the liquid form.

Dipole–Dipole Forces

Dipole–dipole forces occur between neutral, polar molecules: The positive end of one polar molecule is attracted to the negative end of another polar molecule.

Molecules with greater polarity will have greater dipole–dipole attraction, so molecules with larger dipole moments tend to have higher melting and boiling points. Dipole–dipole attractions are relatively weak, however, and these substances melt and boil at very low temperatures. Most substances held together by dipole–dipole attraction are gases or liquids at room temperature.

London Dispersion Forces

London dispersion forces occur between neutral, nonpolar molecules. These very weak attractions occur because of the random motions of electrons on atoms within molecules. At a given moment, a nonpolar molecule might have more electrons on one side than on the other, giving it an instantaneous polarity. For that fleeting instant, the molecule will act as a very weak dipole.

Since London dispersion forces depend on the random motions of electrons, molecules with more electrons will experience greater London dispersion forces. So among substances that experience only London dispersion forces, the one with more electrons will generally have higher melting and boiling points. London dispersion forces are even weaker than dipole–dipole forces, so substances that experience only London dispersion forces melt and boil at extremely low temperatures and tend to be gases at room temperature.

LEWIS DOT STRUCTURES

Drawing Lewis Dot Structures

At some point on the test, you'll be asked to draw the Lewis structure for a molecule or polyatomic ion. Here's how to do it.

1.   Count the valence electrons in the molecule or polyatomic ion; refer to page 24 for a periodic table.
2.   If a polyatomic ion has a negative charge, add electrons equal to the charge of the total in (1). If a polyatomic ion has a positive charge, subtract electrons equal to the charge of the electrons from the total in (1).
3.   Draw the skeletal structure of the molecule and place two electrons (or a single bond) between each pair of bonded atoms. If the molecule contains three or more atoms, the least electronegative atom will usually occupy the central position.
4.   Add electrons to the surrounding atoms until each has a complete outer shell.
5.   Add the remaining electrons to the central atom.
6.   Look at the central atom.
      (a) If the central atom has fewer than eight electrons, remove an electron pair from an outer atom and add another bond between that outer atom and the central atom. Do this until the central atom has a complete octet.
      (b) If the central atom has a complete octet, you are finished.
      (c) If the central atom has more than eight electrons, that's okay too.

Let's find the Lewis structure for the CO_3^{2-} ion.

1.   Carbon has 4 valence electrons; oxygen has 6.
      4 + 6 + 6 + 6 = 22
2.   The ion has a charge of –2, so add 2 electrons.
      22 + 2 = 24
3.   Carbon is the central atom.

4.   Add electrons to the oxygen atoms.

5.   We've added all 24 electrons, so there's nothing left to put on the carbon atom.

6.   (a) We need to give carbon a complete octet, so we take an electron pair away from one of the oxygens and make a double bond instead. Place a bracket around the model and add a charge of negative two.

Resonance Forms

When we put a double bond into the CO_3^{2-} ion, we place it on any one of the oxygen atoms, as shown below.

All three resonance forms are considered to exist simultaneously, and the strength and lengths of all three bonds are the same: somewhere between the strength and length of a single bond and a double bond.

Incomplete Octets

Some atoms can have a complete outer shell with less than eight electrons; for example, hydrogen can have a maximum of two electrons, and beryllium can be stable with only four valence electrons, as in BeH_2.

$$H - Be - H$$

Boron can be stable with only six valence electrons, as in BF_3.

Expanded Octets

In molecules that have *d* subshells available, the central atom can have more than eight valence electrons.

Here are some examples.

PCl₅

SF₄

SF₆

Odd Numbers of Electrons

Molecules almost always have an even number of electrons, allowing electrons to be paired, but there are exceptions, usually involving nitrogen.

NO

Note that NO₂ can be shown with either of two resonance forms.

NO₂

Molecular Geometry

Electrons repel one another, so when atoms come together to form a molecule, the molecule will assume the shape that keeps its different electron pairs as far apart as possible. When we predict the geometries of molecules using this idea, we are using the **valence shell electron-pair repulsion (VSEPR) model.**

In a molecule with more than two atoms, the shape of the molecule is determined by the number of electron pairs on the central atom. The central atom forms **hybrid orbitals,** each of which has a standard shape. Variations on the standard shape occur depending on the number of bonding pairs and lone pairs of electrons on the central atom.

Here are some things you should remember when dealing with the VSEPR model.

- Double and triple bonds are treated in the same way as single bonds in terms of predicting overall geometry for a molecule; however, multiple bonds have slightly more repulsive strength and will therefore occupy a little more space than single bonds.
- Lone electron pairs have a little more repulsive strength than bonding pairs, so lone pairs will occupy a little more space than bonding pairs.

The tables on the following pages show the different hybridizations and geometries that you might see on the test.

If the central atom has **2** electron pairs, then it has *sp* hybridization and its basic shape is **linear.**

Number of lone pairs	Geometry	Examples
0	B—A—B linear	$BeCl_2$ CO_2

If the central atom has **3** electron pairs, then it has sp^2 hybridization and its basic shape is **trigonal planar**; its bond angles are about 120°.

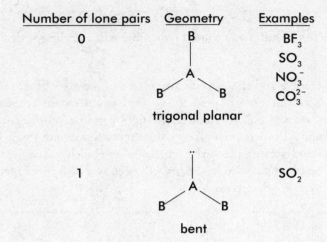

Number of lone pairs	Geometry	Examples
0	trigonal planar	BF_3 SO_3 NO_3^- CO_3^{2-}
1	bent	SO_2

If the central atom has **4** electron pairs, then it has sp^3 hybridization and its basic shape is **tetrahedral**; its bond angles are about 109.5°.

Number of lone pairs	Geometry	Examples
0	tetrahedral	CH_4 NH_4^+ ClO_4^- SO_4^{2-} PO_4^{3-}
1	trigonal pyramidal	NH_3 PCl_3 AsH_3 SO_3^{2-}
2	bent	H_2O OF_2 NH_2^-

If the central atom has **5** electron pairs, then it has *dsp³* hybridization and its basic shape is **trigonal bipyramidal**.

Number of lone pairs	Geometry	Examples

0 — trigonal bipyramidal — PCl_5, PF_5

1 — folded square, seesaw, distorted tetrahedron — SF_4, IF_4^+

2 — T-shaped — ClF_3, ICl_3

3 — linear — XeF_2, I_3^-

In trigonal bipyrimidal shapes, place the lone pairs in axial position first. In octahedral shapes, place lone pairs in equatorial position first.

If the central atom has **6** electron pairs, then it has d^2sp^3 hybridization and its basic shape is **octahedral**.

Number of lone pairs	Geometry	Examples
0		SF_6

octahedral

1		BrF_5 IF_5

square pyramidal

2		XeF_4 ICl_4^-

square planar

Bonding and Phases

The phase of a substance is directly related to the strength of their intermolecular forces. Solids have highly ordered structures where the atoms are packed tightly together, while gases has atoms spread so far apart that most of the volume is free space.

| Solid | Liquid | Gas |

Thus, the strength of the intermolecular force plays a role in the phase behavior of a substance. In other words, substances that exhibit weak interactions (e.g. London Dispersion Forces) tend to be gases at room temperature (e.g. N_2), while substances that exhibit strong interactions (e.g. ionic bonds) tend to be solids at room temperature (e.g. NaCl).

KINETIC MOLECULAR THEORY

For ideal gases, the following assumptions can be made:

- The kinetic energy of an ideal gas is directly proportional to its absolute temperature: The greater the temperature, the greater the average kinetic energy of the gas molecules.

The Total Kinetic Energy of a Gas Sample

$$KE = \frac{3}{2} nRT$$

R = the gas constant; 8.31 joules/mol-K
T = absolute temperature (K)
n = number of moles (mol)

The Average Kinetic Energy of a Single Gas Molecule

$$KE = \frac{1}{2} mv^2$$

m = mass of the molecule (kg)
v = speed of the molecule (meters/sec)
KE is measured in joules

- If several different gases are present in a sample at a given temperature, all the gases will have the same average kinetic energy. That is, the average kinetic energy of a gas depends only on the absolute temperature, not on the identity of the gas.
- The volume of an ideal gas particle is insignificant when compared with the volume in which the gas is contained.
- There are no forces of attraction between the gas molecules in an ideal gas.
- Gas molecules are in constant motion, colliding with one another and with the walls of their container.

THE IDEAL GAS EQUATION

You can use the ideal gas equation to calculate any of the four variables relating to the gas, provided that you already know the other three.

The Ideal Gas Equation

$$PV = nRT$$

P = the pressure of the gas (atm)

V = the volume of the gas (L)

n = the number of moles of gas

T = the absolute temperature of the gas (K)

R = the gas constant, 0.0821 L-atm/mol-K

You can also manipulate the ideal gas equation to figure out how changes in each of its variables affect the other variables.

$$\frac{P_1 V_1}{T_1} = \frac{P_2 V_2}{T_2}$$

P = the pressure of the gas (atm)

V = the volume of the gas (L)

T = the absolute temperature of the gas (K)

You should be comfortable with the following simple relationships:

- If the volume is constant: As pressure increases, temperature increases; as temperature increases, pressure increases.

- If the temperature is constant: As pressure increases, volume decreases; as volume increases, pressure decreases. That's Boyle's law.
- If the pressure is constant: As temperature increases, volume increases; as volume increases, temperature increases. That's Charles's law.

DALTON'S LAW

Dalton's law states that the total pressure of a mixture of gases is just the sum of all the partial pressures of the individual gases in the mixture.

$$\text{Dalton's Law}$$
$$P_{total} = P_a + P_b + P_c + \ldots$$

You should also note that the partial pressure of a gas is directly proportional to the number of moles of that gas present in the mixture. So if 25 percent of the gas in a mixture is helium, then the partial pressure due to helium will be 25 percent of the total pressure.

$$\text{Partial Pressure}$$
$$P_a = (P_{total})(X_a)$$

$$X_a = \frac{\text{moles of gas A}}{\text{total moles of gas}}$$

VAN DER WAALS EQUATION

At low temperature and/or high pressure, gases behave in a less-than-ideal manner. That's because the assumptions made in kinetic molecular theory become invalid under conditions where gas molecules are packed too tightly together.

Two things happen when gas molecules are packed too tightly.

- *The volume of the gas molecules becomes significant.*
  The ideal gas equation does not take the volume of gas molecules into account, so the actual volume of a gas under nonideal conditions will be larger than the volume predicted by the ideal gas equation.
- *Gas molecules attract one another and stick together.*
  The ideal gas equation assumes that gas molecules never stick together. When a gas is packed tightly together, van der Waals forces

(dipole–dipole attractions and London dispersion forces) become significant, causing some gas molecules to stick together. When gas molecules stick together, there are fewer particles bouncing around and creating pressure, so the real pressure in a nonideal situation will be smaller than the pressure predicted by the ideal gas equation.

The van der Waals equation adjusts the ideal gas equation to take nonideal conditions into account.

---

van der Waals Equation

$$\left(P + \frac{n^2 a}{V^2}\right)(V - nb) = nRT$$

P = the pressure of the gas (atm)

V = the volume of the gas (L)

n = the number of moles of gas (mol)

T = the absolute temperature of the gas (K)

R = the gas constant, 0.0821 L-atm/mol-K

a = a constant, different for each gas, that takes into account the attractive forces between molecules

b = a constant, different for each gas, that takes into account the volume of each molecule

---

DENSITY

You may be asked about the density of a gas. The density of a gas is measured in the same way as the density of a liquid or solid: in mass per unit volume.

Density of a Gas

$$D = \frac{m}{V}$$

D = density

m = mass of gas, usually in grams

V = volume occupied by a gas, usually in liters

R = the gas constant, 0.0821 L-atm/mol-K

SOLUTIONS

Molarity

Molarity (*M*) expresses the concentration of a solution in terms of volume. It is the most widely used unit of concentration, turning up in calculations involving equilibrium, acids and bases, and electrochemistry, among others.

When you see a chemical symbol in brackets on the test, that means they are talking about molarity. For instance, "[Na^+]" is the same as "the molar concentration (molarity) of sodium ions."

$$\text{Molarity } (M) = \frac{\text{moles of solute}}{\text{liters of solution}}$$

Molality

Molality (*m*) expresses concentration in terms of the mass of a solvent. It is the unit of concentration used for determining the effect of most colligative properties, where the number of moles of solute is more important than the nature of the solute.

$$\text{Molality } (m) = \frac{\text{moles of solute}}{\text{kilograms of solvent}}$$

Molarity and molality differ in two ways: Molarity tells you about moles of solute per *volume* of the *entire solution* (that is, the solute and the solvent), whereas molality tells you about moles of solute per *mass* of the *solvent*. Keeping in mind that one liter of water weighs one kilogram, and that for a dilute solution, the amount of solution is about the same as the amount of solvent, you should be able to see that for dilute aqueous solutions, molarity and molality are basically the same.

Mole Fraction

Mole fraction (X_s) gives the fraction of moles of a given substance (S) out of the total moles present in a sample. It is used in determining how the vapor pressure of a solution is lowered by the addition of a solute.

$$\text{Mole Fraction } (X_s) = \frac{\text{moles of substance S}}{\text{total number of moles in solution}}$$

Solutes and Solvents

There is a basic rule for remembering which solutes will dissolve in which solvents.

> ### Like dissolves like

That means that polar or ionic solutes (like salt) will dissolve in polar solvents (like water). That also means that nonpolar solutes (like organic compounds) are best dissolved in nonpolar solvents. When an ionic substance dissolves, it breaks up into ions. That's **dissociation.** Free ions in a solution are called electrolytes because they can conduct electricity.

The **van't Hoff factor (i)** tells how many ions one unit of a substance will dissociate into in solution. For instance

- $C_6H_{12}O_6$ does not dissociate, so $i = 1$
- NaCl dissociates into Na^+ and Cl^-, so $i = 2$
- HNO_3 dissociates into H^+ and NO_3^-, so $i = 2$
- $CaCl_2$ dissociates into Ca^{2+}, Cl^-, and Cl^-, so $i = 3$

CHAPTER 4 QUESTIONS

Multiple-Choice Questions

Questions 1–4

    (A) Metallic bonding
    (B) Network covalent bonding
    (C) Hydrogen bonding
    (D) Ionic bonding

1. Solids exhibiting this kind of bonding are excellent conductors of heat.

2. This kind of bonding is the reason that water is more dense than ice.

3. This kind of bonding exists between atoms with very different electronegativities.

4. The stability exhibited by diamonds is due to this kind of bonding.

Questions 5–7

    (A) CH_4
    (B) NH_3
    (C) NaCl
    (D) N_2

5. This substance undergoes ionic bonding.

6. This molecule contains two pi (π) bonds.

7. This substance undergoes hydrogen bonding.

Questions 8–10

    (A) BF_3
    (B) CO_2
    (C) H_2O
    (D) CF_4

8. The central atom in this molecule forms sp^2 hybrid orbitals.

9. This molecule has a tetrahedral structure.

10. This molecule has a linear structure.

11. A liquid whose molecules are held together by which of the following forces would be expected to have the lowest boiling point?

   (A) Ionic bonds
   (B) London dispersion forces
   (C) Hydrogen bonds
   (D) Metallic bonds

12. Which of the following species does NOT have a tetrahedral structure?

   (A) CH_4
   (B) NH_4^+
   (C) SF_4
   (D) $AlCl_4^-$

13. Which form of orbital hybridization can form molecules with shapes that are either trigonal pyramidal or tetrahedral?

   (A) sp
   (B) sp^2
   (C) sp^3
   (D) d^2sp

14. The six carbon atoms in a benzene molecule are shown in different resonance forms as three single bonds and three double bonds. If the length of a single carbon–carbon bond is 154 pm and the length of a double carbon–carbon bond is 133 pm, what length would be expected for the carbon–carbon bonds in benzene?

   (A) 126 pm
   (B) 133 pm
   (C) 140 pm
   (D) 154 pm

15. Which of the following could be the Lewis structure for sulfur trioxide?

   (A)

   (B)

   (C)

   (D)

16. A molecule whose central atom has d^2sp^3 hybridization can have which of the following shapes?

      I. Tetrahedral
     II. Square pyramidal
   III. Square planar

  (A) I only
  (B) III only
  (C) I and II only
  (D) II and III only

17. Which of the following molecules will have a Lewis dot structure with exactly one unshared electron pair on the central atom?

  (A) H_2O
  (B) PH_3
  (C) PCl_5
  (D) CH_2Cl_2

18. Which of the following lists of species is in order of increasing boiling points?

  (A) H_2, N_2, NH_3
  (B) N_2, NH_3, H_2
  (C) NH_3, H_2, N_2
  (D) NH_3, N_2, H_2

19. Which of the compounds listed below would require the greatest energy to separate it into ions in the gaseous state?

  (A) NaCl
  (B) NaI
  (C) MgO
  (D) Na_2O

20. Which of the molecules listed below has the largest dipole moment?

  (A) Cl_2
  (B) HCl
  (C) SO_3
  (D) NO

21. The temperature of a sample of an ideal gas confined in a 2.0 L container was raised from 27°C to 77°C. If the initial pressure of the gas was 1,200 mmHg, what was the final pressure of the gas?

  (A)   300 mmHg
  (B)   600 mmHg
  (C) 1,400 mmHg
  (D) 2,400 mmHg

22. A sealed container containing 8.0 grams of oxygen gas and 7.0 of nitrogen gas is kept at a constant temperature and pressure. Which of the following is true?

  (A) The volume occupied by oxygen is greater than the volume occupied by nitrogen.
  (B) The volume occupied by oxygen is equal to the volume occupied by nitrogen.
  (C) The volume occupied by nitrogen is greater than the volume occupied by oxygen.
  (D) The density of nitrogen is greater than the density of oxygen.

23. A mixture of gases contains 1.5 moles of oxygen, 3.0 moles of nitrogen, and 0.5 moles of water vapor. If the total pressure is 700 mmHg, what is the partial pressure of the nitrogen gas?

(A) 210 mmHg
(B) 280 mmHg
(C) 350 mmHg
(D) 420 mmHg

24. A mixture of helium and neon gases has a total pressure of 1.2 atm. If the mixture contains twice as many moles of helium as neon, what is the partial pressure due to neon?

(A) 0.2 atm
(B) 0.3 atm
(C) 0.4 atm
(D) 0.8 atm

25. Nitrogen gas was collected over water at 25°C. If the vapor pressure of water at 25°C is 23 mmHg, and the total pressure in the container is measured at 781 mmHg, what is the partial pressure of the nitrogen gas?

(A)   46 mmHg
(B) 551 mmHg
(C) 735 mmHg
(D) 758 mmHg

26. A gas sample is confined in a 5-liter container. Which of the following will occur if the temperature of the container is increased?

  I.   The kinetic energy of the gas will increase.
  II. The pressure of the gas will increase.
  III. The density of the gas will increase.

(A) I only
(B) II only
(C) I and II only
(D) I, II, and III

27. A 22.0 gram sample of an unknown gas occupies 11.2 liters at standard temperature and pressure. Which of the following could be the identity of the gas?

(A) CO_2
(B) SO_3
(C) O_2
(D) He

28. Which of the following conditions would be most likely to cause the ideal gas laws to fail?

  I.   High pressure
  II. High temperature
  III. Large volume

(A) I only
(B) II only
(C) I and II only
(D) II and III only

29. Which of the following expressions is equal to the density of helium gas at standard temperature and pressure?

(A) $\dfrac{1}{22.4}$ g/L

(B) $\dfrac{2}{22.4}$ g/L

(C) $\dfrac{1}{4}$ g/L

(D) $\dfrac{4}{22.4}$ g/L

30. In an experiment $H_2(g)$ and $O_2(g)$ were completely reacted, above the boiling point of water, according to the following equation in a sealed container of constant volume and temperature:

$$2H_2(g) + O_2(g) \rightarrow 2H_2O(g)$$

If the initial pressure in the container before the reaction is denoted as P_i, which of the following expressions gives the final pressure, assuming ideal gas behavior?

(A) P_i
(B) $2\,P_i$
(C) $(3/2)P_i$
(D) $(2/3)P_i$

31. Nitrogen gas was collected over water at a temperature of 40°C, and the pressure of the sample was measured at 796 mmHg. If the vapor pressure of water at 40°C is 55 mmHg, what is the partial pressure of the nitrogen gas?

(A) 741 mmHg
(B) 756 mmHg
(C) 796 mmHg
(D) 851 mmHg

32. An ideal gas fills a balloon at a temperature of 27°C and 1 atm pressure. By what factor will the volume of the balloon change if the gas in the balloon is heated to 127°C at constant pressure?

(A) $\dfrac{27}{127}$

(B) $\dfrac{3}{4}$

(C) $\dfrac{4}{3}$

(D) $\dfrac{2}{1}$

33. A gas sample with a mass of 10 grams occupies 6.0 liters and exerts a pressure of 2.0 atm at a temperature of 26°C. Which of the following expressions is equal to the molecular mass of the gas? The gas constant, R, is 0.08 (L-atm)/(mol-K).

(A) $\dfrac{(10)(0.08)(299)}{(2.0)(6.0)}$ g/mol

(B) $\dfrac{(299)(0.08)}{(10)(2.0)(6.0)}$ g/mol

(C) $\dfrac{(2.0)(6.0)(299)}{(10)(0.08)}$ g/mol

(D) $\dfrac{(10)(2.0)(6.0)}{(299)(0.08)}$ g/mol

34. Which of the following assumptions is (are) valid based on kinetic molecular theory?

   I. Gas molecules have negligible volume.
   II. Gas molecules exert no attractive forces on one another.
   III. The temperature of a gas is directly proportional to its kinetic energy.

(A) I only
(B) III only
(C) I and III only
(D) I, II, and III

35. A substance is dissolved in water, forming a 0.50-molar solution. If 4.0 liters of solution contains 240 grams of the substance, what is the molecular mass of the substance?

(A)   60 grams/mole
(B) 120 grams/mole
(C) 240 grams/mole
(D) 480 grams/mole

36. A solution contains equal masses of glucose (molecular mass 180) and toluene (molecular mass 90). What is the mole fraction of glucose in the solution?

(A) $\dfrac{1}{4}$

(B) $\dfrac{1}{3}$

(C) $\dfrac{1}{2}$

(D) $\dfrac{2}{3}$

37. How many moles of Na_2SO_4 must be added to 500 milliliters of water to create a solution that has a 2-molar concentration of the Na^+ ion? (Assume the volume of the solution does not change).

(A) 0.5 moles
(B) 1 mole
(C) 2 moles
(D) 5 moles

38. Given that a solution of NaCl (molar mass 58.5 g/mole) in water (molar mass 18 g/mole) has a molality of 0.5 m, which of the following can be determined?

   I. The mass of the NaCl in the solution
   II. The total mass of the solution
   III. The mole fraction of the NaCl in the solution

   (A) I only
   (B) III only
   (C) I and II only
   (D) II and III only

Free Response Questions

1. Use the principles of bonding and molecular structure to explain the following statements.

   (a) The boiling point of argon is $-186°C$, whereas the boiling point of neon is $-246°C$.
   (b) Solid sodium melts at $98°C$, but solid potassium melts at $64°C$.
   (c) More energy is required to break up a $CaO(s)$ crystal into ions than to break up a $KF(s)$ crystal into ions.
   (d) Molten KF conducts electricity, but solid KF does not.

2. The carbonate ion CO_3^{2-} is formed when carbon dioxide, CO_2, reacts with slightly basic cold water.

   (a) (i) Draw the Lewis electron dot structure for the carbonate ion. Include resonance forms if they apply.
       (ii) Draw the Lewis electron dot structure for carbon dioxide.
   (b) Describe the hybridization of carbon in the carbonate ion.
   (c) (i) Describe the relative lengths of the three C–O bonds in the carbonate ion.
       (ii) Compare the average length of the C–O bonds in the carbonate ion to the average length of the C–O bonds in carbon dioxide.

3.

Substance	Boiling Point (°C)	Bond Length (Å)	Bond Strength (kcal/mol)
H_2	−253	0.75	104.2
N_2	−196	1.10	226.8
O_2	−182	1.21	118.9
Cl_2	−34	1.99	58.0

   (a) Explain the differences in the properties given in the table above for each of the following pairs.
       (i) The bond strengths of N_2 and O_2
       (ii) The bond lengths of H_2 and Cl_2
       (iii) The boiling points of O_2 and Cl_2
   (b) Use the principles of molecular bonding to explain why H_2 and O_2 are gases at room temperature, while H_2O is a liquid at room temperature.

4. H_2S, SO_4^{2-}, XeF_2, ICl_4^-

    (a)   Draw a Lewis electron dot diagram for each of the molecules listed above.
    (b)   Use the valence shell electron-pair repulsion (VSEPR) model to predict the geometry of each of the molecules.

5. Use the principles of bonding and molecular structure to explain the following statements.

    (a)   The angle between the N–F bonds in NF_3 is smaller than the angle between the B–F bonds in BF_3.
    (b) $I_2(s)$ is insoluble in water, but it is soluble in carbon tetrachloride.
    (c)   Diamond is one of the hardest substances on Earth.
    (d)   HCl has a lower boiling point than either HF or HBr.

6.

The graph above shows the changes in pressure with changing temperature of gas samples of helium and argon confined in a closed 2-liter vessel.

    (a)   What is the total pressure of the two gases in the container at a temperature of 200 K?
    (b)   How many moles of helium are contained in the vessel?
    (c)   How many molecules of helium are contained in the vessel?
    (d)   What is the ratio of the average speeds of the helium atoms to the average speeds of the argon atoms?
    (e)   If the volume of the container were reduced to 1 liter at a constant temperature of 300 K, what would be the new pressure of the helium gas?

7. $2\ KClO_3(s) \rightarrow 2\ KCl(s) + 3\ O_2(g)$

The reaction above took place, and 1.45 liters of oxygen gas were collected over water at a temperature of 29°C and a pressure of 755 millimeters of mercury. The vapor pressure of water at 29°C is 30.0 millimeters of mercury.

    (a)   What is the partial pressure of the oxygen gas collected?
    (b)   How many moles of oxygen gas were collected?
    (c)   What would be the dry volume of the oxygen gas at a pressure of 760 millimeters of mercury and a temperature of 273 K?
    (d)   What was the mass of the $KClO_3$ consumed in the reaction?

8. Equal molar quantities of two gases, O_2 and H_2O, are confined in a closed vessel at constant temperature.

    (a)   Which gas, if any, has the greater partial pressure?
    (b)   Which gas, if any, has the greater density?
    (c)   Which gas, if any, has the greater concentration?
    (d)   Which gas, if any, has the greater average kinetic energy?
    (e)   Which gas, if any, will show the greater deviation from ideal behavior?

9. SF_6, H_2O, and CO_2 are all greenhouse gases because they absorb infrared radiation in the atmosphere. As such they are of interest to many researchers.

    (a)   Of the three gases, which deviates the most from ideal gas behavior?
    (b)   If all three ideal gases are held at constant temperature, which gas would have the highest average molecular speed?
    (c)   A container was with equal amounts H_2O and CO_2 and the pressure was measured. Assuming ideal behavior, if SF_6 was then added such that the final pressure was four times that before its addition, the final number of moles of SF_6 is how many times that of H_2O?
    (d)   If a container held equal molar amounts of all three greenhouse gases, but was then compressed decreasing the volume within, which of the gases has the highest, final partial pressure assuming ideal behavior?
    (e)   Gas molecules exert pressure through collisions with the walls of the containers they are in. If a container is filled with equal amounts of each of the greenhouse gases, which gas undergoes the fewest collisions with the wall of the container per unit time?

CHAPTER 4 ANSWERS AND EXPLANATIONS

Multiple-Choice Questions

1. **A**   In metallic bonding, nuclei are surrounded by a sea of mobile electrons. The electrons' freedom to move allows them to conduct heat and electricity.

2. **C**   When ice forms, the hydrogen bonds join the molecules in a lattice structure, which forces them to remain farther apart than they were in the liquid form. Because the molecules are farther apart in the solid than in the liquid, the solid (ice) is less dense than the liquid.

3. **D**   Electronegativity is a measure of how much pull an atom exerts on another atom's electrons. If the difference in electronegativities is large enough (greater than 1.7) then the more electronegative atom will simply take an electron away from the other atom. The two atoms will then be held together by electrostatic attraction (the atom that has gained an electron becomes negative and the atom that has lost an electron becomes positive). That's an ionic bond.

4. **B**   The carbon atoms in diamond are held together by a network of covalent bonds. The carbon atoms form sp^3 hybrid orbitals, resulting in a tetrahedral structure, which is very stable and has no simple breaking points.

5. **C**   NaCl is composed of two elements that have very different electronegativities than each other, so Na gives up an electron to Cl and the two are held together by electrostatic attraction in an ionic bond.

6. **D** N_2 contains a triple bond, so it has one sigma (σ) bond and two pi (π) bonds.

7. **B**   Hydrogen bonding occurs between hydrogen atoms of one molecule and electronegative elements (F, O, or N) of another molecule. So in ammonia, hydrogens from one ammonia molecule will form bonds with nitrogens from another ammonia molecule.

8. **A**   In BF_3, boron forms three bonds with fluorine atoms and has no unbonded valence electrons, so it must form sp^2 hybrid orbitals.

9. **D** CF_4 forms a tetrahedral structure as shown in the diagram below. The central carbon atom is hybridized sp^3.

10. **B** CO_2 forms a linear structure as shown in the diagram below. The central carbon atom is sp hybridized.

$$\ddot{O}=C=\ddot{O}$$

11. **B**   A liquid with a low boiling point must be held together by weak bonds. London dispersion forces are the weakest kind of intermolecular forces.

12. **C** SF_4 has 34 valence electrons distributed in the Lewis dot structure and shape shown below.

In this molecule, sulfur forms dsp^3 hybrid orbitals, which have a trigonal bipyramid structure. Because SF_4 has one unshared electron pair, the molecule takes the "seesaw" or "folded square" shape.

Choices (A) and (B), CH_4 and NH_4^+, each have 8 valence electrons distributed in the same Lewis dot structure and shape, shown below for NH_4^+.

In these molecules, the central atom forms sp^3 hybrid orbitals, which have a tetrahedral structure. There are no unshared electron pairs on the central atom, so the molecules are tetrahedral.

Choice (D) $AlCl_4^-$ has 32 valence electrons distributed in the same Lewis dot structure and shape, shown below for $AlCl_4^-$.

In these molecules, the central atom forms sp^3 hybrid orbitals, which have a tetrahedral structure. There are no unshared electron pairs on the central atom, so the molecules are tetrahedral.

13. **C** The sp^3 hybrid orbitals take a tetrahedral shape if the central atom has no unshared electron pairs (CH_4, for instance). If the central atom has one unshared electron pair, the molecule takes the trigonal pyramid shape (NH_3, for instance).

14. **C** Resonance is used to describe a situation that lies between single and double bonds, so the bond length would also be expected to be in between that of single and double bonds.

15. **A** Choice (A) has the correct number of valence electrons (6 + 6 + 6 + 4 + 2 = 24), and 8 valence electrons on each atom.

About the other choices:

(B) There are only 6 valence electrons on the sulfur atom.

(C) There are too many valence electrons (10) on the sulfur atom.

(D) There are too many valence electrons (26).

16. **D** A molecule with d^2sp^3 hybridization has octahedral structure if the central atom has no unbonded electrons (SF_6, for instance).

If the central atom has one unbonded electron pair, the molecule is square pyramidal (IF_5, for instance).

If the central atom has two unbonded electron pairs, the molecule is square planar (XeF_4, for instance).

A molecule with d^2sp^3 hybridization can never be tetrahedral.

17. **B** The Lewis dot structures for the answer choices are shown below. Only PH_3 has a single unshared electron pair on its central atom.

(A)

(B)

(C)

(D)

(E)

18.   **A** H_2 experiences only van der Waals forces and has the lowest boiling point.

N₂ also experiences only van der Waals forces, but it is larger than H_2 and has more electrons, so it has stronger van der Waals interactions with other molecules.

NH_3 is polar and undergoes hydrogen bonding, so it has the strongest intermolecular interactions and the highest boiling point.

19.   **C**   The energy required to separate an ionic compound into gaseous ions is called the lattice energy. The most important factor in determining the lattice energy is the strength of the charges on the two ions. In MgO, the Mg ion is +2 and the O ion is –2. All of the other choices listed contain +1 or –1 ions, so their lattice energies will not be as large.

20.   **B**   The bond that holds HCl together is a covalent bond with a large polarity. The bond that holds NO together is also polar covalent, but its polarity is very small because N and O are so close together on the periodic table. Cl_2 and N_2 are nonpolar because they share electrons equally, and SO_3 is nonpolar because it is symmetrical (trigonal planar). Nonpolar molecules have dipole moments of zero.

21.   **C**   Remember to convert Celsius to Kelvin (°C + 273 = K).

27°C = 300 K and 77°C = 350 K.

From the relationship $\dfrac{P_1}{T_1} = \dfrac{P_2}{T_2}$ we get

$$\frac{\left(1,200 \text{ mm Hg}\right)}{\left(300 \text{ K}\right)} = \frac{P_2}{\left(350 \text{ K}\right)}$$

So P_2 = 1,400 mmHg

22.   **B**   8.0 grams of oxygen and 7.0 grams of nitrogen both equal 0.25 moles. Avogadro's law states equal volumes contain equal moles. This makes (A) and (C) wrong.

(D) This is reversed: The density of oxygen gas would be greater than the density of nitrogen.

23. **D** From Dalton's law, the partial pressure of a gas depends on the number of moles of the gas that are present.

The total number of moles of gas present is

$$1.5 + 3.0 + 0.5 = 5.0 \text{ total moles}$$

If there are 3 moles of nitrogen, then $\frac{3}{5}$ of the pressure must be due to nitrogen.

$$(\frac{3}{5})(700 \text{ mmHg}) = 420 \text{ mmHg}$$

24. **C** From Dalton's law, the partial pressure of a gas depends on the number of moles of the gas that are present. If the mixture has twice as many moles of helium as neon, then the mixture must be $\frac{1}{3}$ neon. So $\frac{1}{3}$ of the pressure must be due to neon.

$$(\frac{1}{3})(1.2 \text{ atm}) = 0.4 \text{ atm.}$$

25. **D** From Dalton's law, the partial pressures of nitrogen and water vapor must add up to the total pressure in the container. The partial pressure of water vapor in a closed container will be equal to the vapor pressure of water, so the partial pressure of nitrogen is

$$781 \text{ mmHg} - 23 \text{ mmHg} = 758 \text{ mmHg}$$

26. **C** From kinetic molecular theory, we know that kinetic energy is directly proportional to temperature (think of the expression $KE = \frac{3}{2}kT$), so (I) is true.

From the gas laws, we know that at constant volume, an increase in temperature will bring about an increase in pressure (think of $\frac{P_1}{T_1} = \frac{P_2}{T_2}$), so (II) is true.

The density of a gas is equal to mass per unit volume, which is not changed by changing temperature, so (III) is not true.

27. **A** Use the following relationship:

$$\text{Moles} = \frac{\text{liters}}{22.4 \text{ L/mol}}$$

$$\text{Moles of unknown gas} = \frac{11.2 \text{ L}}{22.4 \text{ L/mol}} = 0.500 \text{ moles}$$

$$MW = \frac{\text{grams}}{\text{mole}}$$

$$MW \text{ of unknown gas} = \frac{22.0 \text{ g}}{0.500 \text{ mole}} = 44.0 \text{ grams/mole}$$

That's the molecular weight of CO_2.

28. **A** The ideal gas laws fail under conditions where gas molecules are packed too tightly together. This can happen as a result of high pressure (I), low temperature (not listed as a choice), or small volume (not listed as a choice).

29. **D** Density is measured in grams per liter. One mole of helium gas has a mass of 4 grams and occupies a volume of 22.4 liters at STP, so the density of helium gas at STP is $\frac{4}{22.4}$ g/L.

30. **D** As long as the gases act ideally, the total pressure in the container will be a function only of the number of gas moles present so long as volume and temperature are held constant. The relationship between moles and pressure before and after the reaction, from the ideal gas equation ($PV = nRT$), will be $P_i/n_i = P_f/n_f$ where n_i, P_f, n_f are the initial moles, the final pressure, and the final moles respectively. We can rearrange this equation to $P_f = P_i(n_f/n_i)$, and since there are 2 moles of gas in the products for every 3 moles in the reactants, we can say that $P_f = (2/3)P_i$.

31. **A** From Dalton's law, the partial pressures of nitrogen and water vapor must add up to the total pressure in the container. The partial pressure of water vapor in a closed container will be equal to its vapor pressure, so the partial pressure of nitrogen is

796 mmHg – 55 mmHg = 741 mmHg

32. **C** From the ideal gas laws, for a gas sample at constant pressure

$$\frac{V_1}{T_1} = \frac{V_2}{T_2}$$

Solving for V_2 we get $V_2 = V_1 \dfrac{T_2}{T_1}$

So V_1 is multiplied by a factor of $\dfrac{T_2}{T_1}$

Remember to convert Celsius to Kelvin, $\dfrac{127\,°C + 273}{27\,°C + 273} = \dfrac{400\text{ K}}{300\text{ K}} = \dfrac{4}{3}$

33. **A** We can find the number of moles of gas from $PV = nRT$.

Remember to convert 26°C to 299 K.

Then solve for n.

$$n = \frac{PV}{RT} = \frac{(2.0)(6.0)}{(0.08)(299)}\text{ mol}$$

Now, remember:

$$MW = \frac{\text{grams}}{\text{moles}} = \frac{(10\text{ g})}{\left(\dfrac{(2.0)(6.0)}{(0.08)(299)}\text{ mol}\right)} = \frac{(10)(0.08)(299)}{(2.0)(6.0)}\text{ g/mol}$$

34. **D** All three assumptions are included in kinetic molecular theory.

35. **B** First find the number of moles.

Moles = (molarity)(volume)
Moles of substance = (0.50 M)(4.0 L) = 2 moles

$$\text{Moles} = \frac{\text{grams}}{MW}$$

So $MW = \dfrac{240\text{ g}}{2\text{ mol}}$ = 120 g/mol

36. **B** Let's say the solution contains 180 grams of glucose and 180 grams of toluene. That's 1 mole of glucose and 2 moles of toluene. So that's 1 mole of glucose out of a total of 3 moles, for a mole fraction of $\dfrac{1}{3}$.

37. **A**   Let's find out how many moles of Na^+ we have to add.

Moles = (molarity)(volume)
Moles of Na^+ = (2 M)(0.5 L) = 1 mole

Because we get 2 moles of Na^+ ions for every mole of Na_2SO_4 we add, we need to add only 0.5 moles of Na_2SO_4.

38. **B**   We can't determine (I) and (II) because we don't know how much solution we have. We can figure out (III) because molality tells us the number of moles of NaCl found in 1 kilogram of water. We can figure out how many moles of water there are in 1 kilogram. So if we have a ratio of moles of NaCl to moles of water, we can figure out the mole fraction of NaCl.

Free Response Questions

1. (a)   Molecules of noble gases in the liquid phase are held together by London dispersion forces, which are weak interactions brought about by instantaneous polarities in nonpolar atoms and molecules.

Atoms with more electrons are more easily polarized and experience stronger London dispersion forces. Argon has more electrons than neon, so it experiences stronger London dispersion forces and boils at a higher temperature.

(b)   Sodium and potassium are held together by metallic bonds and positively charged ions in a delocalized sea of electrons.

Potassium is larger than sodium, so the electrostatic attractions that hold the atoms together act at a greater distance, reducing the attractive force and resulting in its lower melting point.

(c)   Both CaO(s) and KF(s) are held together by ionic bonds in crystal lattices.

Ionic bonds are held together by an electrostatic force, which can be determined by using Coulomb's law.

$$F = k\frac{Q_1 Q_2}{r^2}$$

CaO is more highly charged, with Ca^{2+} bonded to O^{2-}. So for CaO, Q_1 and Q_2 are +2 and −2.

KF is not as highly charged, with K^+ bonded to F^-. So for KF, Q_1 and Q_2 are +1 and −1.

CaO is held together by stronger forces and is more difficult to break apart.

(d)   KF is composed of K^+ and F^- ions. In the liquid (molten) state, these ions are free to move and can thus conduct electricity.

In the solid state, the K^+ and F^- ions are fixed in a crystal lattice and their electrons are localized around them, so there is no charge that is free to move and thus no conduction of electricity.

2.   (a)   (i)

(ii)

(b)   The central carbon atom forms three sigma bonds with oxygen atoms and has no free electron pairs, so its hybridization must be sp^2.

(c)   (i)   All three bonds will be the same length because no particular resonance form is preferred over the others. The actual structure is an average of the resonance structures.

(ii) The C–O bonds in the carbonate ion have resonance forms between single and double bonds, while the C–O bonds in carbon dioxide are both double bonds.

The bonds in the carbonate ion will be shorter than single bonds and longer than double bonds, so the carbonate bonds will be longer than the carbon dioxide bonds.

3. (a) (i) The bond strength of N_2 is larger than the bond strength of O_2 because N_2 molecules have triple bonds and O_2 molecules have double bonds. Triple bonds are stronger and shorter than double bonds.

(ii) The bond length of H_2 is smaller than the bond length of Cl_2 because hydrogen is a smaller atom than chlorine, allowing the hydrogen nuclei to be closer together.

(iii) Liquid oxygen and liquid chlorine are both nonpolar substances that experience only London dispersion forces of attraction. These forces are greater for Cl_2 because it has more electrons (which makes it more polarizable), so Cl_2 has a higher boiling point than O_2.

(b) H_2 and O_2 are both nonpolar molecules that experience only London dispersion forces, which are too weak to form the bonds required for a substance to be liquid at room temperature.

H_2O is a polar substance whose molecules form hydrogen bonds with each other. Hydrogen bonds are strong enough to form the bonds required in a liquid at room temperature.

4. (a)

(b) H_2S has two bonds and two free electron pairs on the central S atom. The greatest distance between the electron pairs is achieved by tetrahedral arrangement. The electron pairs at two of the four corners will cause the molecule to have a bent shape, like water.

SO_4^{2-} has four bonds around the central S atom and no free electron pairs. The four bonded pairs will be farthest apart when they are arranged in a tetrahedral shape, so the molecule is tetrahedral.

XeF_2 has two bonds and three free electron pairs on the central Xe atom. The greatest distance between the electron pairs can be achieved by a trigonal bipyramidal arrangement. The three free electron pairs will occupy the equatorial positions, which are 120 degrees apart, to minimize repulsion. The two F atoms are at the poles, so the molecule is linear.

ICl_4^- has four bonds and two free electron pairs on the central I atom. The greatest distance between the electron pairs can be achieved by an octahedral arrangement. The two free electron pairs will be opposite each other to minimize repulsion. The four Cl atoms are in the equatorial positions, so the molecule is square planar.

5.   (a) BF_3 has three bonds on the central B atom and no free electron pairs, so the structure of BF_3 is trigonal planar, with each of the bonds 120 degrees apart.

NF$_3$ has three bonds and one free electron pair on the central N atom. The four electron pairs are pointed toward the corners of a tetrahedron, 109.5 degrees apart. The added repulsion from the free electron pair causes the N–F bonds to be even closer together, and the angle between them is more like 107°.

(b)   Polar solvents are best at dissolving polar solutes. Nonpolar solvents are best at dissolving nonpolar solutes.

$I_2(s)$ is nonpolar, so it dissolves better in carbon tetrachloride, CCl_4, which is nonpolar, than in water, H_2O, which is polar.

(c)   The carbon atoms in diamond are bonded together in a tetrahedral network, with each carbon atom bonded to three other carbon atoms. The tetrahedral structure of the network bonds does not leave any seams along which the diamond can be broken, so a diamond behaves as one big molecule with no weaknesses.

(d) HBr and HCl are polar molecules. In liquid form, both substances are held together by dipole interactions. These interactions are stronger for molecules with more electrons, so HBr has stronger intermolecular bonds and a higher boiling point.

HF has a higher boiling point than HCl because HF undergoes hydrogen bonding, while HCl does not; this causes HF to remain a liquid at higher temperatures than HCl, although HF is a polar molecule with fewer electrons than HCl.

6. (a) Read the graph, and add the two pressures.

$P_{Total} = P_{He} + P_{Ar}$

P_{Total} = (1 atm) + (1.5 atm) = 2.5 atm

(b) Read the pressure (1 atm) at 200 K, and use the ideal gas equation.

$$n = \frac{PV}{RT} = \frac{(1.0 \text{ atm})(2.0 \text{ L})}{(0.082 \text{ L-atm/mol-K})(200 \text{ K})} = 0.12 \text{ moles}$$

(c) Use the definition of a mole.

Molecules = (moles)(6.02 × 10²³)

Molecules (atoms) of helium = (0.12)(6.02 × 10²³) = 7.2 × 10²²

(d) Use Graham's law.

$$\frac{v_1}{v_2} = \sqrt{\frac{MW_2}{MW_1}}$$

$$\frac{v_{He}}{v_{Ar}} = \sqrt{\frac{MW_{Ar}}{MW_{He}}} = \sqrt{\frac{(40)}{(4)}} = 3.2 \text{ to } 1$$

(e) Use the relationship

$$\frac{P_1 V_1}{T_1} = \frac{P_2 V_2}{T_2}$$

With T constant

$$P_1 V_1 = P_2 V_2$$

$$(1.5 \text{ atm})(2.0 \text{ L}) = P_2(1.0 \text{ L})$$

$$P_2 = 3.0 \text{ atm}$$

7. (a) Use Dalton's law.

$$P_{Total} = P_{Oxygen} + P_{Water}$$

$$(755 \text{ mmHg}) = (P_{Oxygen}) + (30.0 \text{ mmHg})$$

$$P_{Oxygen} = 725 \text{ mmHg}$$

(b) Use the ideal gas law. Don't forget to convert to the proper units.

$$n = \frac{PV}{RT} = \frac{\left(\frac{725}{760} \text{ atm}\right)(1.45 \text{ L})}{\left(0.082 \text{ L-atm/mol-K}\right)(302 \text{ K})} = 0.056 \text{ moles}$$

(c) At STP, moles of gas and volume are directly related.

Volume = (moles)(22.4 L/mol)

Volume of O_2 = (0.056 mol)(22.4 L/mol) = 1.25 L

(d) We know that 0.056 moles of O_2 were produced in the reaction.

From the balanced equation, we know that for every 3 moles of O_2 produced, 2 moles of $KClO_3$ are consumed. So there are $\frac{2}{3}$ as many moles of $KClO_3$ as O_2.

Moles of $KClO_3$ = $(\frac{2}{3})$(moles of O_2)

Moles of $KClO_3$ = $(\frac{2}{3})$(0.056 mol) = 0.037 moles

Grams = (moles)(MW)

Grams of $KClO_3$ = (0.037 mol)(122 g/mol) = 4.51 g

8. (a) The partial pressures depend on the number of moles of gas present. Because the number of moles of the two gases are the same, the partial pressures are the same.

(b) O_2 has the greater density. Density is mass per unit volume. Both gases have the same number of moles in the same volume, but oxygen has heavier molecules, so it has greater density.

(c) Concentration is moles per volume. Both gases have the same number of moles in the same volume, so their concentrations are the same.

(d) According to kinetic-molecular theory, the average kinetic energy of a gas depends only on the temperature. Both gases are at the same temperature, so they have the same average kinetic energy.

(e) H_2O will deviate most from ideal behavior. Ideal behavior for gas molecules assumes that there will be no intermolecular interactions.

H_2O is polar, and O_2 is not. H_2O undergoes hydrogen bonding while O_2 does not. So H_2O has stronger intermolecular interactions, which will cause it to deviate more from ideal behavior.

9. (a) H_2O deviates most drastically from ideal behavior due to very strong hydrogen bonding interactions. A truly ideal gas has no interactions with surrounding molecules. These hydrogen bonding interactions are why H_2O is a liquid at room temperature and the others are gases.

(b)   As per Graham's law, the species with the lowest molecular weight would have the greatest average molecular speed. H_2O, MW = 18 g/mol, is the lightest, and hence would be the fastest.

(c)   By Dalton's law of partial pressures, if SF_6 was added until the pressure was 4 times the pressure with just (an equal mixture of) H_2O and CO_2, then there must be 3 times as much SF_6 as H_2O and CO_2 combined. This means that, in the end, there 6 times more SF_6 than H_2O. For example, imagine starting with 1 atm each of CO_2 and H_2O. In this case the initial pressure is 2 atm, and to get to 8 atm one must add 6 atm of SF_6 which is 6 times that of the H_2O present.

(d)   All three gases will have identical partial pressures. The equation for partial pressure of any component a is $P_a = (P_{total})(X_a)$. Since we know that each has the same mol fraction the total pressure doesn't matter, and each will have the same partial pressure, relative to one another.

(e) SF_6 will have the fewest collisions per unit time. Since the kinetic theory assumes that the gas molecules are moving randomly, the one with the most collisions with the walls of the containers will be the one moving the fastest. On the other hand, the slowest molecule will have the fewest collisions. The slowest molecule will be the most massive one, as per Graham's law, which in this case is SF_6.

Chapter 5
Big Idea #3: Chemical Reactions, Energy Changes, and Redox Reactions

Changes in matter involve the rearrangement and/or reorganization of atoms and/or the transfer of electrons

TYPES OF REACTIONS

Reactions may be classified into several categories.

1. Synthesis Reactions – When simple compounds are combined to form a single, more complex compound.

    $$2\,Mg(s) + O_2(g) \rightarrow 2\,MgO(s)$$

2. Decomposition – The opposite of a synthesis. A reaction where a single compound is split into two or more simple compounds, usually in the presence of heat.

    $$HgO(s) + Heat \rightarrow Hg(s) + \tfrac{1}{2}\,O_2(g)$$

3. Acid-Base Reaction – A reaction when an acid (i.e. H^+) reacts with a base (i.e. OH^-) to form water and a salt.

    $$HCl(aq) + NaOH(aq) \rightarrow NaCl(aq) + H_2O(l)$$

4. Oxidation-Reduction (Redox) Reaction – A reaction that results in the change of the oxidation states of some participating molecules.

    $$Cu^{2+}(aq) + 2e^- \rightarrow Cu(s)$$

CHEMICAL EQUATIONS

Balancing Chemical Equations

Normally, balancing a chemical equation is a trial-and-error process. You start with the most complicated-looking compound in the equation and work from there. There is, however, an old Princeton Review SAT trick that you may want to try if you see a balancing equation question on the multiple-choice section. The trick is called **backsolving**.

It works like this: To make a balancing equation question work in a multiple-choice format, one of the answer choices is the correct coefficient for one of the species in the reaction. So instead of starting blind in the trial-and-error process, you can insert the answer choices one by one to see which one works. You probably won't have to try all five, and if you start in the middle and the number doesn't work, it might be obviously too small or large, eliminating other choices before you have to try them. Let's try it.

$$\ldots NH_3 + \ldots O_2 \rightarrow \ldots N_2 + \ldots H_2O$$

1. If the equation above were balanced with lowest whole number coefficients, the coefficient for NH_3 would be

   (A) 1
   (B) 2
   (C) 3
   (D) 4

Here's How to Crack It!

Start at (C) because it's the middle number. If there are 3 NH_3's, then there can't be a whole number coefficient for N_2, so (C) is wrong, and so is the other odd number answer, (A).

Try (D).

If there are 4 NH_3's, then there must be 2 N_2's and 6 H_2O's.

If there are 6 H_2O's, then there must be 3 O_2's, and the equation is balanced with lowest whole number coefficients.

Backsolving is more efficient than the methods that you're accustomed to. If you use the answer choices that you're given, you streamline the trial-and-error process and allow yourself to use POE as you work on the problem.

Chemical Equations and Calculations

Many of the stoichiometry problems on the test will be formatted in the following way: You will be given a balanced chemical equation and told that you have some number of grams (or liters of gas, or molar concentration, and so on) of reactant. Then you will be asked what number of grams (or liters of gas, or molar concentration, and so on) of products are generated.

In these cases, follow this simple series of steps.

1. Convert whatever quantity you are given into moles.
2. If you are given information about two reactants, you may have to use the equation coefficients to determine which one is the limiting reagent. Remember: The limiting reagent is not necessarily the reactant that you have the least of; it is the reactant that runs out first.
3. Use the balanced equation to determine how many moles of the desired product are generated.
4. Convert moles of product to the desired unit.

Let's try one.

$$2\ HBr(aq) + Zn(s) \rightarrow ZnBr_2(aq) + H_2(g)$$

2. A piece of solid zinc weighing 98 grams was added to a solution containing 324 grams of HBr. What is the volume of H_2 produced at standard temperature and pressure if the reaction above runs to completion?

   (A) 11 liters
   (B) 22 liters
   (C) 34 liters
   (D) 45 liters

Here's How to Crack It!

1. Convert whatever quantity you are given into moles.

$$\text{Moles of Zn} = \frac{\text{grams}}{\text{molar mass}} = \frac{(98\ g)}{(65.4\ g/mol)} = 1.5\ mol$$

$$\text{Moles of HBr} = \frac{\text{grams}}{\text{molar mass}} = \frac{(324\ g)}{(80.9\ g/mol)} = 4.0\ mol$$

2. Use the balanced equation to find the limiting reagent.
    From the balanced equation, 2 moles of HBr are used for every mole of Zn that reacts, so when 1.5 moles of Zn react, 3 moles of HBr are consumed, and there will be HBr left over when all of the Zn is gone. That makes Zn the limiting reagent.

3. Use the balanced equation to determine how many moles of the desired product are generated.
    1 mole of H_2 is produced for every mole of Zn consumed, so if 1.5 moles of Zn are consumed, then 1.5 moles of H_2 are produced.

4. Convert moles of product to the desired unit.
    The H_2 gas is at STP, so we can convert directly from moles to volume.
    Volume of H_2 = (moles)(22.4 L/mol) = (1.5 mol)(22.4 L/mol) = 33.6 L $\approx$ 34 L

So (C) is correct.

Let's try another one using the same reaction.

———————————⚬———————————

$$2 \, HBr(aq) + Zn(s) \rightarrow ZnBr_2(aq) + H_2(g)$$

3. A piece of solid zinc weighing 13.1 grams was placed in a container. A 0.10-molar solution of HBr was slowly added to the container until the zinc was completely dissolved. What was the volume of HBr solution required to completely dissolve the solid zinc?

(A) 1.0 L
(B) 2.0 L
(C) 3.0 L
(D) 4.0 L

Here's How to Crack It!

1. Convert whatever quantity you are given into moles.

$$\text{Moles of Zn} = \frac{\text{grams}}{\text{MW}} = \frac{(13.1 \, \text{g})}{(65.4 \, \text{g/mol})} = 0.200 \, \text{mol}$$

2. Use the balanced equation to find the limiting reagent.

3. Use the balanced equation to determine how many moles of the desired product are generated.

In this case, we're using the balanced reaction to find out how much of one reactant is required to consume the other reactant. It's a slight variation on the process described in (2) and (3).

We can see from the balanced equation that it takes 2 moles of HBr to react completely with 1 mole of Zn, so it will take 0.400 moles of HBr to react completely with 0.200 moles of Zn.

4. Convert moles of product to the desired unit.

Moles of HBr = (molarity)(volume)

$$\text{Volume of HBr} = \frac{\text{moles}}{\text{molarity}} = \frac{(0.400 \, \text{mol})}{(0.10 \, \text{mol/L})} = 4.0 \, \text{L}$$

So (D) is correct.

———————————⚬———————————

When you perform calculations, always include units. Including units in your calculations will help you (and the person scoring your test) keep track of what you are doing. Including units will also get you partial credit points on the free-response section.

ENTHALPY

Enthalpy Change, ΔH

The enthalpy of a substance is a measure of the energy that is released or absorbed by the substance when bonds are broken and formed during a reaction.

> **The Basic Rules of Enthalpy**
> When bonds are *formed*, energy is *released*.
> When bonds are *broken*, energy is *absorbed*.

The change in enthalpy, ΔH, that takes place over the course of a reaction can be calculated by subtracting the enthalpy of the reactants from the enthalpy of the products.

> **Enthalpy Change**
> $$\Delta H = H_{products} - H_{reactants}$$

All substances like to be in the lowest possible energy state, which gives them the greatest stability. This means that, in general, exothermic processes are more likely to occur spontaneously than endothermic processes.

If the products have stronger bonds than the reactants, then the products have lower enthalpy than the reactants and are more stable; in this case, energy is released by the reaction, or the reaction is **exothermic**.

If the products have weaker bonds than the reactants, then the products have higher enthalpy than the reactants and are less stable; in this case, energy is absorbed by the reaction, or the reaction is **endothermic**.

ENERGY DIAGRAMS

Exothermic and Endothermic Reactions

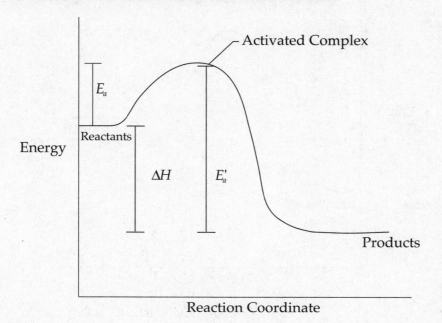

EXOTHERMIC REACTION

The diagram above shows the energy change that takes place during an exothermic reaction. The reactants start with a certain amount of energy (read the graph from left to right). For the reaction to proceed, the reactants must have enough energy to reach the transition state, where they are part of an activated complex. This is the highest point on the graph above. The amount of energy needed to reach this point is called the **activation energy, E_a**. At this point, all reactant bonds have been broken, but no product bonds have been formed, so this is the point in the reaction with the highest energy and lowest stability. The energy needed for the reverse reaction is shown as line E_a'.

Moving to the right past the activated complex, product bonds start to form, and we eventually reach the energy level of the products.

This diagram represents an exothermic reaction, so the products are at a lower energy level than the reactants and ΔH is negative.

The diagram below shows an endothermic reaction.

ENDOTHERMIC REACTION

In this diagram, the energy of the products is greater than the energy of the reactants, so ΔH is positive.

Reaction diagrams can be read in both directions, so the reverse reaction for an exothermic reaction is endothermic and vice versa.

CATALYSTS AND ENERGY DIAGRAMS

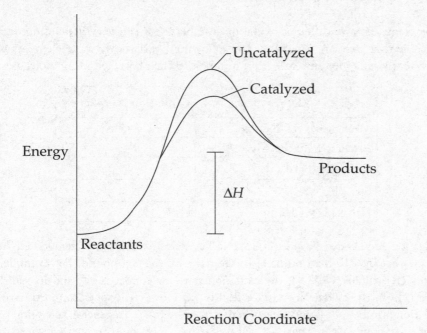

A catalyst speeds up a reaction by providing the reactants with an alternate pathway that has a lower activation energy, as shown in the diagram above.

Notice that the only difference between the catalyzed reaction and the uncatalyzed reaction is that the energy of the activated complex is lower for the catalyzed reaction. A catalyst lowers the activation energy, but it has no effect on the energy of the reactants, the energy of the products, or ΔH for the reaction.

Also note that a catalyst lowers the activation energy for both the forward and the reverse reaction, so it has no effect on the equilibrium conditions.

OXIDATION STATES

The **oxidation state** (or oxidation number) of an atom indicates the number of electrons that it gains or loses when it forms a bond. For instance, upon forming a bond with another atom, oxygen generally gains two electrons, which are negatively charged, so the oxidation state of oxygen in a bond is −2. Similarly, sodium generally loses one electron when it bonds to another atom, so its oxidation state in a bond is +1. Here are three important things you have to keep in mind when dealing with oxidation numbers.

- The oxidation state of an atom that is not bonded to an atom of another element is zero. That means either an atom that is not bonded to any other atom or an atom that is bonded to another atom of the same element (like the oxygen atoms in O_2).
- The oxidation numbers for all the atoms in a molecule must add up to zero.

- The oxidation numbers for all the atoms in a polyatomic ion must add up to the charge on the ion.

Most elements have different oxidation numbers that can vary depending on the molecule that they are a part of. The following chart shows some elements that consistently take the same oxidation numbers as their corresponding molecules.

Element	Oxidation Number
Alkali metals (Li, Na, ...)	+1
Alkaline earths (Be, Mg, ...)	+2
Group 3A (B, Al, ...)	+3
Oxygen	−2
Halogens (F, Cl, ...)	−1

Transition metals can have several oxidation states, which are differentiated from one another by a Roman numeral in the name of the compound. For example, in copper (II) sulfate ($CuSO_4$), the oxidation state for copper is +2, and in lead (II) oxide (PbO), the oxidation state for lead is +2. However, copper can also have an oxidation number of +1, and Pb can range from −4 to +4. In general, transition metals or d-block metals are characterized by variable or changing oxidation states.

You should be familiar with the following polyatomic ions and their charges.

Hydroxide	OH^-
Nitrate	NO_3^-
Perchlorate	ClO_4^-
Acetate	$C_2H_3O_2^-$
Carbonate	CO_3^{2-}
Sulfate	SO_4^{2-}
Phosphate	PO_4^{3-}

OXIDATION-REDUCTION REACTIONS

In an oxidation-reduction (or redox, for short) reaction, electrons are exchanged by the reactants, and the oxidation states of some of the reactants are changed over the course of the reaction. Look at the following reaction:

$$Fe + 2\ HCl \rightarrow FeCl_2 + H_2$$

The oxidation state of Fe changes from 0 to +2.

The oxidation state of H changes from +1 to 0.

- *When an atom gains electrons, its oxidation number decreases, and it is said to have been reduced.*

In the reaction above, H was reduced.

- *When an atom loses electrons, its oxidation number increases, and it is said to have been oxidized.*

In the reaction above, Fe was oxidized.

Oxidation and reduction go hand in hand. If one atom is losing electrons, another atom must be gaining them.

- If an atom is losing electrons and being oxidized, it must be giving the electrons to another atom, which is being reduced. So if a reactant contains an atom that is being oxidized, it is a **reducing agent** or **reductant**.
- If an atom is taking electrons and being reduced, it must be taking electrons away from another atom, which is being oxidized. So if a reactant contains an atom that is being reduced, it is an **oxidizing agent** or **oxidant**.

An oxidation-reduction reaction can be written as two **half-reactions**: one for the reduction and one for the oxidation. For example, the reaction

$$Fe + 2 \; HCl \rightarrow FeCl_2 + H_2$$

can be written as

$$Fe \rightarrow Fe^{2+} + 2 \; e^- \qquad \text{Oxidation}$$
$$2 \; H^+ + 2 \; e^- \rightarrow H_2 \qquad \text{Reduction}$$

Reduction Potentials

Every half-reaction has a potential, or voltage, associated with it. For Section II of the test, you'll be given a table of standard reduction potentials like the one seen on the next page. The potentials are given as reduction half-reactions, but you can read them in reverse and change the sign on the voltage to get oxidation potentials.

> Here's a mnemonic device that might be useful.
>
> LEO the lion says GER
> **LEO:** you Lose Electrons in Oxidation
> **GER:** you Gain Electrons in Reduction

Standard Reduction Potentials, E, in Water Solution at 25°C (in V)			
$F_2(g) + 2 e^-$	$\rightarrow$	$2 F^-$	2.87
$Co^{3+} + e^-$	$\rightarrow$	Co^{2+}	1.82
$Au^{3+} + 3e^-$	$\rightarrow$	$Au(s)$	1.50
$Cl_2(g) + 2 e^-$	$\rightarrow$	$2 Cl^-$	1.36
$O_2(g) + 4H^+ + 4 e^-$	$\rightarrow$	$2 H_2O$	1.23
$Br_2(l) + 2 e^-$	$\rightarrow$	$2 Br^-$	1.07
$2 Hg^{2+} + 2 e^-$	$\rightarrow$	Hg_2^{2+}	0.92
$Hg^{2+} + 2 e^-$	$\rightarrow$	$Hg(l)$	0.85
$Ag^+ + e^-$	$\rightarrow$	$Ag(s)$	0.80
$Hg_2^{2+} + 2 e^-$	$\rightarrow$	$2 Hg(l)$	0.79
$Fe^{3+} + e^-$	$\rightarrow$	Fe^{2+}	0.77
$I_2(s) + 2 e^-$	$\rightarrow$	$2 I^-$	0.53
$Cu^+ + e^-$	$\rightarrow$	$Cu(s)$	0.52
$Cu^{2+} + 2 e^-$	$\rightarrow$	$Cu(s)$	0.34
$Cu^{2+} + e^-$	$\rightarrow$	Cu^+	0.15
$Sn^{4+} + 2 e^-$	$\rightarrow$	Sn^{2+}	0.15
$S(s) + 2 H^+ + 2 e^-$	$\rightarrow$	H_2S	0.14
$2 H^+ + 2 e^-$	$\rightarrow$	$H_2(g)$	0.00
$Pb^{2+} + 2 e^-$	$\rightarrow$	$Pb(s)$	−0.13
$Sn^{2+} + 2 e^-$	$\rightarrow$	$Sn(s)$	−0.14
$Ni^{2+} + 2 e^-$	$\rightarrow$	$Ni(s)$	−0.25
$Co^{2+} + 2 e^-$	$\rightarrow$	$Co(s)$	−0.28
$Tl^+ + e^-$	$\rightarrow$	$Tl(s)$	−0.34
$Cd^{2+} + 2 e^-$	$\rightarrow$	$Cd(s)$	−0.40
$Cr^{3+} + e^-$	$\rightarrow$	Cr^{2+}	−0.41
$Fe^2 + 2 e^-$	$\rightarrow$	$Fe(s)$	−0.44
$Cr^{3+} + 3 e^-$	$\rightarrow$	$Cr(s)$	−0.74
$Zn^{2+} + 2 e^-$	$\rightarrow$	$Zn(s)$	−0.76
$Mn^{2+} + 2 e^-$	$\rightarrow$	$Mn(s)$	−1.18
$Al^{3+} + 3 e^-$	$\rightarrow$	$Al(s)$	−1.66
$Be^{2+} + 2 e^-$	$\rightarrow$	$Be(s)$	−1.70
$Mg^{2+} + 2 e^-$	$\rightarrow$	$Mg(s)$	−2.37
$Na^+ + e^-$	$\rightarrow$	$Na(s)$	−2.71
$Ca^{2+} + 2 e^-$	$\rightarrow$	$Ca(s)$	−2.87
$Sr^{2+} + 2 e^-$	$\rightarrow$	$Sr(s)$	−2.89
$Ba^{2+} + 2 e^-$	$\rightarrow$	$Ba(s)$	−2.90
$Rb^+ + e^-$	$\rightarrow$	$Rb(s)$	−2.92
$K^+ + e^-$	$\rightarrow$	$K(s)$	−2.92
$Cs^+ + e^-$	$\rightarrow$	$Cs(s)$	−2.92
$Li^+ + e^-$	$\rightarrow$	$Li(s)$	−3.05

Look at the reduction potential for Zn^{2+}.

$$Zn^{2+} + 2\ e^- \rightarrow Zn \qquad E^\circ = -0.76\ V$$

Read the reduction half-reaction in reverse and change the sign on the voltage to get the oxidation potential for Zn.

$$Zn \rightarrow Zn^{2+} + 2\ e^- \qquad E^\circ = +0.76\ V$$

The larger the potential for a half-reaction, the more likely it is to occur; for instance, let's look at the top of the table of half-reactions.

$$F_2(g) + 2e^- \rightarrow 2\ F^- \qquad E^\circ = +2.87\ V$$

$F_2(g)$ has a very large reduction potential, so it is likely to gain electrons and be reduced, making it a strong oxidizing agent.

Now let's look at the bottom of the table. We need to look at the reverse (oxidation) reaction to get a positive potential.

$$Li(s) \rightarrow Li^+ + e^- \qquad E^\circ = +3.05\ V$$

$Li(s)$ has a very large oxidation potential, making it very likely to lose electrons and be oxidized, so $Li(s)$ is a very strong reducing agent.

You can calculate the potential of a redox reaction if you know the potentials for the two half-reactions that constitute it. There are two important things to remember when calculating the potential of a redox reaction.

- Add the potential for the oxidation half-reaction to the potential for the reduction half-reaction.
- Never multiply the potential for a half-reaction by a coefficient.

Let's look at the following reaction:

$$Zn + 2\ Ag^+ \rightarrow Zn^{2+} + 2\ Ag$$

The two half-reactions are

Oxidation:	$Zn \rightarrow Zn^{2+} + 2\ e^-$	$E^\circ = +0.76\ V$
Reduction:	$Ag^+ + e^- \rightarrow Ag$	$E^\circ = +0.80\ V$

$$E = E_{oxidation} + E_{reduction}$$
$$E = 0.76\ V + 0.80\ V = 1.56\ V$$

Notice that we ignored that silver has a coefficient of 2 in the balanced equation.

GALVANIC CELLS

In a **galvanic cell** (also called a voltaic cell), a spontaneous redox reaction is used to generate a flow of current.

Look at the following spontaneous redox reaction:

$$Zn(s) + Cu^{2+}(aq) \rightarrow Zn^{2+}(aq) + Cu(s) \qquad E^{\circ} = 1.10 \text{ V}$$

Oxidation:	$Zn(s) \rightarrow Zn^{2+}(aq) + 2 \text{ e}^-$	$E^{\circ} = 0.76$ V
Reduction:	$Cu^{2+}(aq) + 2 \text{ e}^- \rightarrow Cu(s)$	$E^{\circ} = 0.34$ V

A galvanic cell using this reaction is shown below.

In a galvanic cell, the two half-reactions take place in separate chambers, and the electrons that are released by the oxidation reaction pass through a wire to the chamber where they are consumed in the reduction reaction. That's how the current is created. **Current** is defined as the flow of positive charge, so current is always in the opposite direction from the flow of electrons.

> There's a mnemonic device to remember that.
>
> AN OX
> RED CAT

In any electric cell (either a galvanic cell or an electrolytic cell, which we'll discuss a little later) oxidation takes place at the electrode called the **anode**. Reduction takes place at the electrode called the **cathode**.

The salt bridge maintains electrical neutrality in the system by providing enough negative ions to equal the positive ions being created at the anode (during oxidation) and providing positive ions to replace the Cu^{2+} ions being used up at the cathode (during reduction). The salt bridge can be an actual salt, or it can be a slim passage that allows ions to move between the two chambers.

Under standard conditions (when all concentrations are 1 M) the voltage of the cell is the same as the total voltage of the redox reaction. Under nonstandard conditions, the cell voltage can be computed by using the Nernst equation.

The Nernst Equation

$$E_{cell} = E°_{cell} - \frac{RT}{nF} \ln Q$$

E_{cell} = cell potential under nonstandard conditions (V)

$E°_{cell}$ = cell potential under standard conditions (V)

R = the gas constant, 8.31 (volt-coulomb)/(mol-K)

T = absolute temperature (K)

n = the number of moles of electrons exchanged in the reaction (mol)

F = Faraday's constant, 96,500 coulombs/mole

Q = the reaction quotient (same as the equilibrium expression, but with initial concentrations instead of equilibrium concentrations)

At 25°C, the Nernst equation reduces to

$$E_{cell} = E°_{cell} - \frac{0.0592}{n} \log Q$$

Here's the most important thing you should understand from the Nernst equation: *As the concentration of the products of a redox reaction increases, the voltage decreases; and as the concentration of the reactants in a redox reaction increases, the voltage increases.* This is a direct result of Le Châtelier's law.

ELECTROLYTIC CELLS

In an electrolytic cell, an outside source of voltage is used to force a nonspontaneous redox reaction to take place. Let's look at the electrolysis of molten NaCl.

$2\ Na^+(molten) + 2\ Cl^-(molten) \rightarrow 2\ Na(l) + Cl_2(g)$ $E° = -4.07\ V$

Oxidation: $2\ Cl^-(molten) \rightarrow Cl_2(g) + 2\ e^-$ $E° = -1.36\ V$

Reduction: $2\ Na^+(molten) + 2\ e^- \rightarrow 2\ Na(l)$ $E° = -2.71\ V$

An electrolytic cell that forces this reaction to take place is shown below.

In this process, pure liquid sodium and pure chlorine gas are generated from molten sodium chloride. Notice that the process does not take place in an aqueous solution. That's because water is more easily reduced than Na^+, so water would be reduced instead of sodium ions in an aqueous solution.

The AN OX/RED CAT rule applies to the electrolytic cell in the same way that it applies to the galvanic cell.

Electrolytic cells are used for electroplating. You may see a question on the test that gives you an electrical current and asks you how much metal "plates out."

There are roughly four steps for figuring out electrolysis problems.

1. If you know the current and the time, you can calculate the charge in coulombs.

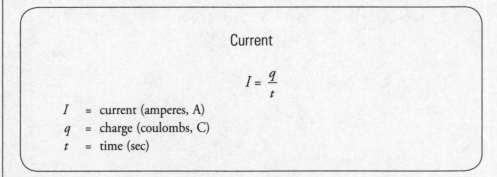

Current

$$I = \frac{q}{t}$$

I = current (amperes, A)
q = charge (coulombs, C)
t = time (sec)

2. Once you know the charge in coulombs, you know how many electrons were involved in the reaction.

$$\text{Moles of electrons} = \frac{\text{coulombs}}{96,500 \text{ coulombs/mol}}$$

3. When you know the number of moles of electrons and you know the half-reaction for the metal, you can find out how many moles of metal plated out. For example, from this half-reaction for gold

$$Au^{3+} + 3 \, e^- \rightarrow Au(s)$$

you know that for every 3 moles of electrons consumed, you get 1 mole of gold.

4. Once you know the number of moles of metal, you can use what you know from stoichiometry to calculate the number of grams of metal.

CHAPTER 5 QUESTIONS

Multiple-Choice Questions

Questions 1-3

    (A) MnO_4^-
    (B) H^+
    (C) H_2
    (D) Na

1. This is a very strong reducing agent.

2. By definition, the reduction potential for this species is equal to zero.

3. This is a very strong oxidizing agent.

Questions 4-6

The choices listed below refer to n, the number of moles of electrons transferred in a reaction.

    (A) $n = 4$
    (B) $n = 3$
    (C) $n = 2$
    (D) $n = 1$

4. $2\,Fe^{3+} + Mg \rightarrow 2\,Fe^{2+} + Mg^{2+}$

5. $F_2 + 2\,Br^- \rightarrow 2\,F^- + Br_2$

6. $MnO_4^- + Cr + 2\,H_2O \rightarrow MnO_2 + Cr^{3+} + 4OH^-$

7. What is the mass ratio of fluorine to boron in a boron trifluoride molecule?
    (A) 1.8 to 1
    (B) 3.0 to 1
    (C) 3.5 to 1
    (D) 5.3 to 1

8. A hydrocarbon sample with a mass of 6 grams underwent combustion, producing 11 grams of carbon dioxide. If all of the carbon initially present in the compound was converted to carbon dioxide, what was the percent of carbon, by mass, in the hydrocarbon sample?
    (A) 25%
    (B) 33%
    (C) 50%
    (D) 66%

9. What is the mass of oxygen in 148 grams of calcium hydroxide ($Ca(OH)_2$)?

   (A) 24 grams
   (B) 32 grams
   (C) 48 grams
   (D) 64 grams

10. An ion containing only oxygen and chlorine is 31% oxygen by mass. What is its empirical formula?

    (A) ClO^-
    (B) ClO_2^-
    (C) ClO_3^-
    (D) ClO_4^-

11. A sample of propane, C_3H_8, was completely burned in air at STP. The reaction occurred as shown below.

$$C_3H_8 + O_2 \rightarrow 3\ CO_2 + 4\ H_2O$$

If 67.2 liters of CO_2 were produced and all of the carbon in the CO_2 came from the propane, what was the mass of the propane sample?

    (A) 11 grams
    (B) 22 grams
    (C) 33 grams
    (D) 44 grams

12. What is the percent composition by mass of the elements in the compound $NaNO_3$?

    (A) Na 20%, N 20%, O 60%
    (B) Na 23%, N 14%, O 48%
    (C) Na 23%, N 14%, O 63%
    (D) Na 27%, N 16%, O 57%

13. The composition of a typical glass used in bottles is 12.0% Na_2O, 12.0% CaO, and 76.0% SiO_2. Which of the following lists the three compounds in order of greatest to least number of moles present in a typical sample of bottle glass?

    (A) SiO_2, CaO, Na_2O
    (B) SiO_2, Na_2O, CaO
    (C) Na_2O, SiO_2, CaO
    (D) Na_2O, CaO, SiO_2

14. The concentration of sodium chloride in sea water is about 0.5 molar. How many grams of NaCl are present in 1 kg of sea water?

    (A) 30 grams
    (B) 60 grams
    (C) 100 grams
    (D) 300 grams

15. A sample of a hydrate of $CuSO_4$ with a mass of 250 grams was heated until all the water was removed. The sample was then weighed and found to have a mass of 160 grams. What is the formula for the hydrate?

    (A) $CuSO_4 \cdot 10\ H_2O$
    (B) $CuSO_4 \cdot 7\ H_2O$
    (C) $CuSO_4 \cdot 5\ H_2O$
    (D) $CuSO_4 \cdot 2\ H_2O$

16. A compound containing only sulfur and oxygen is 50% sulfur by weight. What is the empirical formula for the compound?

    (A) SO
    (B) SO_2
    (C) SO_3
    (D) S_2O

17. $...CN^- + ...OH^- \rightarrow ...CNO^- + ...H_2O + ...e^-$

    When the half reaction above is balanced, what is the coefficient for OH^- if all the coefficients are reduced to the lowest whole number?

    (A) 1
    (B) 2
    (C) 3
    (D) 4

18. $CaCO_3(s) + 2 H+(aq) \rightarrow Ca_2+(aq) + H_2O(l) + CO_2(g)$

    If the reaction above took place at standard temperature and pressure and 150 grams of $CaCO_3(s)$ were consumed, what was the volume of $CO_2(g)$ produced at STP?

    (A) 11 L
    (B) 22 L
    (C) 34 L
    (D) 45 L

19. A gaseous mixture at 25°C contained 1 mole of CH_4 and 2 moles of O_2 and the pressure was measured at 2 atm. The gases then underwent the reaction shown below.

    $$CH_4(g) + 2 O_2(g) \rightarrow CO_2(g) + 2 H_2O(g)$$

    What was the pressure in the container after the reaction had gone to completion and the temperature was allowed to return to 25°C?

    (A) 1 atm
    (B) 2 atm
    (C) 3 atm
    (D) 4 atm

20. A hydrocarbon was found to be 20% hydrogen by weight. If 1 mole of the hydrocarbon has a mass of 30 grams, what is its molecular formula?

    (A) CH_2
    (B) CH_3
    (C) C_2H_4
    (D) C_2H_6

21. $...CuFeS_2 + ...O_2 \rightarrow Cu_2S + ...FeO + ...SO_2$

    When the half-reaction above is balanced, what is the coefficient for O_2 if all the coefficients are reduced to the lowest whole number?

    (A) 2
    (B) 3
    (C) 4
    (D) 6

22. When chlorine gas is combined with fluorine gas, a compound is formed that is 38% chlorine and 62% fluorine. What is the empirical formula of the compound?

(A) ClF
(B) ClF_2
(C) ClF_3
(D) ClF_5

23.
$$Cr_2O_7{}^{2-} + 6I^- + 14\ H^+ \rightarrow 2Cr^{3+} + 3I_2 + 7H_2O$$

Which of the following statements about the reaction given above is NOT true?

(A) The oxidation number of chromium changes from +6 to +3.
(B) The oxidation number of iodine changes from −1 to 0.
(C) The oxidation number of hydrogen changes from +1 to 0.
(D) The oxidation number of oxygen remains the same.

24. $Al^{3+} + 3e^- \rightarrow Al(s)$ $\qquad\qquad$ $E° = -1.66\ V$

$Cr^{3+} + 3e^- \rightarrow Cr(s)$ $\qquad\qquad$ $E° = -0.74\ V$

The standard reduction potentials for two half-reactions are shown above. Which of the statements listed below will be true for the following reaction taking place under standard conditions?

$$Al(s) + Cr^{3+} \rightarrow Al^{3+} + Cr(s)$$

(A) $E° = 2.40\ V$, and the reaction is not spontaneous.
(B) $E° = 0.92\ V$, and the reaction is spontaneous.
(C) $E° = -0.92\ V$, and the reaction is not spontaneous.
(D) $E° = -0.92\ V$, and the reaction is spontaneous.

25. In which of the following molecules does hydrogen have an oxidation state of −1?

(A) H_2O
(B) NH_3
(C) CaH_2
(D) CH_4

26. Oxygen takes the oxidation state −1 in hydrogen peroxide, H_2O_2. The equation for the decomposition of H_2O_2 is shown below.

$$2\ H_2O_2 \rightarrow 2\ H_2O + O_2$$

Which of the following statements about the reaction shown above is true?

(A) Oxygen is reduced, and hydrogen is oxidized.
(B) Oxygen is oxidized, and hydrogen is reduced.
(C) Oxygen is both oxidized and reduced.
(D) Hydrogen is both oxidized and reduced.

27. When solid iron is brought into contact with water and oxygen, it undergoes the following half-reaction:

$$Fe(s) \rightarrow Fe^{2+}(aq) + 2e^-$$

This half-reaction is instrumental in the corrosion of iron. When iron is coated with solid zinc, the half-reaction above is impeded, even if the zinc coating is incomplete. This is most likely because

(A) $Zn(s)$ is more easily reduced than $Fe(s)$.
(B) $Zn(s)$ is more easily oxidized than $Fe(s)$.
(C) $Zn^{2+}(aq)$ is more easily reduced than $Fe(s)$.
(D) $Zn^{2+}(aq)$ is more easily oxidized than $Fe(s)$.

28. $2 H_2O(l) + 2e^- \rightarrow H_2(g) + 2 OH^-(aq)$ $E° = -0.8$ V

   $Na^+(aq) + e^- \rightarrow Na(s)$ $E° = -2.7$ V

   $Cl_2(g) + 2e^- \rightarrow Cl^-(aq)$ $E° = +1.4$ V

Based on the reduction potentials given above, which of the following would be expected to occur when electrodes connected to the terminals of a 2.0 V battery are immersed in a solution of sodium chloride in water?

(A) Solid sodium will appear at the anode, and chlorine gas will appear at the cathode.
(B) Chlorine gas will appear at the anode, and solid sodium will appear at the cathode.
(C) Hydrogen gas will appear at the anode, and chlorine gas will appear at the cathode.
(D) Chlorine gas will appear at the anode, and hydrogen gas will appear at the cathode.

29. $Cu^{2+} + 2e^- \rightarrow Cu$ $E° = +0.3$ V

   $Fe^{2+} + 2e^- \rightarrow Fe$ $E° = -0.4$ V

Based on the reduction potentials given above, what is the reaction potential for the following reaction?

$$Fe^{2+} + Cu \rightarrow Fe + Cu^{2+}$$

(A) −0.7 V
(B) −0.1 V
(C) +0.1 V
(D) +0.7 V

30. $Cu^{2+} + 2e^- \rightarrow Cu \qquad E° = +0.3$ V
$Zn^{2+} + 2e^- \rightarrow Zn \qquad E° = -0.8$ V
$Mn^{2+} + 2e^- \rightarrow Mn \qquad E° = -1.2$ V

Based on the reduction potentials given above, which of the following reactions will occur spontaneously?

(A) $Mn^{2+} + Cu \rightarrow Mn + Cu^{2+}$
(B) $Mn^{2+} + Zn \rightarrow Mn + Zn^{2+}$
(C) $Zn^{2+} + Cu \rightarrow Zn + Cu^{2+}$
(D) $Zn^{2+} + Mn \rightarrow Zn + Mn^{2+}$

31. $F_2 + 2e^- \rightarrow 2F^- \qquad E° = 2.87$ V
$Ag^{2+} + e^- \rightarrow Ag^+ \qquad E° = 1.99$ V
$Au^{3+} + 3e^- \rightarrow 3Au° \qquad E° = 1.50$ V
$Cl_2 + 2e^- \rightarrow 2Cl^- \qquad E° = 1.36$ V
$3Ag^+ + 3e^- \rightarrow 3Ag° \qquad E° = 0.80$ V

In two separate experiments, pieces of metallic gold and metallic silver are reacted. In the first experiment both are reacted an inert solvent with Cl_2, while in the second they are reacted with F_2. Which of the following represents the metallic species present after the second reaction, which were not present after the first?

(A) Ag^+ only
(B) Ag^{2+} and Au^+
(C) Ag^+ and Ag^{2+}
(D) Ag^{2+} and Au^{3+}

32. Many metallic ions in the +1 oxidation state (generically stated, M^{+1}) are known to undergo *disproportionation* reactions in which two M^{+1} ions electrochemically react to form one M^{2+} and one $M°$. If this reaction for the generic metal M^+ is spontaneous at room temperature, which of the following 1 electron standard potentials *must* sum to a positive number?

(A) the reduction potential of M^{2+} and the reduction potential of M^+
(B the reduction potential of M^+ and the oxidation potential of M^+
(C) the oxidation potential of M^{2+} and the oxidation potential of M^+
(D) the oxidation potential of $M°$ and the reduction potential of M^{2+}

33. Molten $AlCl_3$ is electrolyzed with a constant current of 5.00 amperes over a period of 600.0 seconds. Which of the following expressions is equal to the maximum mass of $Al(s)$ that plates out? (1 faraday = 96,500 coulombs)

(A) $\dfrac{(600)(5.00)}{(96,500)(3)(27.0)}$ grams

(B) $\dfrac{(600)(5.00)(3)(27.0)}{(96,500)}$ grams

(C) $\dfrac{(600)(5.00)(27.0)}{(96,500)(3)}$ grams

(D) $\dfrac{(96,500)(3)(27.0)}{(600)(5.00)}$ grams

Free Response Questions

1. A 10.0 gram sample containing calcium carbonate and an inert material was placed in excess hydrochloric acid. A reaction occurred producing calcium chloride, water, and carbon dioxide.

    (a) Write the balanced equation for the reaction.
    (b) When the reaction was complete, 900 milliliters of carbon dioxide gas were collected at 740 mmHg and 30°C. How many moles of calcium carbonate were consumed in the reaction?
    (c) If all of the calcium carbonate initially present in the sample was consumed in the reaction, what percent by mass of the sample was due to calcium carbonate?
    (d) If the inert material was silicon dioxide, what was the molar ratio of calcium carbonate to silicon dioxide in the original sample?

2. A gaseous hydrocarbon sample is completely burned in air, producing 1.80 liters of carbon dioxide at standard temperature and pressure and 2.16 grams of water.

    (a) What is the empirical formula for the hydrocarbon?
    (b) What was the mass of the hydrocarbon consumed?
    (c) The hydrocarbon was initially contained in a closed 1.00 liter vessel at a temperature of 32°C and a pressure of 760 millimeters of mercury. What is the molecular formula of the hydrocarbon?
    (d) Write the balanced equation for the combustion of the hydrocarbon.

3. The table below shows three common forms of copper ore.

Ore #	Empirical Formula	Percent by Weight		
		Copper	Sulfur	Iron
1	Cu_2S	?	?	0
2	?	34.6	34.9	30.5
3	?	55.6	28.1	16.3

    (a) What is the percent by weight of copper in Cu_2S?
    (b) What is the empirical formula of ore #2?
    (c) If a sample of ore #3 contains 11.0 grams of iron, how many grams of sulfur does it contain?
    (d) Cu can be extracted from Cu_2S by the following process:
$$3\ Cu_2S + 3\ O_2 \rightarrow 3\ SO_2 + 6\ Cu$$
    If 3.84 grams of O_2 are consumed in the process, how many grams of Cu are produced?

4. $$2\ Mg(s) + 2\ CuSO_4(aq) + H_2O(l) \rightarrow 2\ MgSO_4(aq) + Cu_2O(s) + H_2(g)$$

    (a) If 1.46 grams of $Mg(s)$ are added to 500 milliliters of a 0.200-molar solution of $CuSO_4$, what is the maximum molar yield of $H_2(g)$?
    (b) When all of the limiting reagent has been consumed in (a), how many moles of the other reactant (not water) remain?
    (c) What is the mass of the Cu_2O produced in (a)?
    (d) What is the value of $[Mg^{2+}]$ in the solution at the end of the experiment? (Assume that the volume of the solution remains unchanged.)

5. An 800-milliliter sample of 0.080-molar Ag^+ solution was electrolyzed, resulting in the formation of solid silver and oxygen gas. The solution was subjected to a current of 2.00 amperes for 10.0 minutes. The solution became progressively more acidic as the reaction progressed.

   (a) Write the two half-reactions that occur, stating which takes place at the anode and which at the cathode.
   (b) If the oxygen gas produced in the electrolysis was collected at standard temperature and pressure, what was its volume?
   (c) What was the mass of solid silver produced in the electrolysis?
   (d) If the solution was neutral at the start of the electrolysis, what was the pH of the solution when the process was completed?

6. Use the principles of electrochemistry to answer each of the following:

   (a) Explain why a salt bridge connecting the two compartments of a galvanic cell is necessary for the operation of the cell.
   (b) Explain why, when an iron nail is placed in hydrochloric acid, a reaction occurs, but when a copper penny is placed in hydrochloric acid, no reaction occurs.
   (c) $Ag^+ + Sn(s) \rightarrow Ag(s) + Sn^{2+}$

      (i) Give the standard cell potential for the reaction above.
      (ii) What happens to the cell potential in (i) when $[Ag^+]$ is increased?
      (iii) What happens to the cell potential when the amount of $Sn(s)$ in the cell is increased?

7. $$M(s) + Cd^{2+} \rightarrow M^+ + Cd(s)$$

   The reaction above proceeds spontaneously at 25°C in an electrochemical cell.

   (a) The reduction potential for M(s) must be less than a certain value. What is that value?
   (b) Write the two half-reactions, and identify which takes place at the anode and which takes place at the cathode.
   (c) Describe the changes that occur in the quantities below as the reaction proceeds in the forward direction.
      (i) $[M^+]$
      (ii) E_{cell}
      (iii) G

8. Molten sodium chloride can be electrolyzed to form pure sodium and chlorine.

   (a) Write the balanced equations for the half-reactions that take place at the electrodes during the electrolysis. Indicate which half-reaction takes place at which electrode.
   (b) Describe how the pure sodium and chlorine are separated.
   (c) For the electrolyzation of molten sodium chloride, will the value of each of the following be positive or negative?
      (i) $E_{reaction}$
      (ii) ΔG
      (iii) ΔS
   (d) Explain why, when the molten sodium chloride is replaced with an aqueous solution of sodium chloride, pure sodium is not produced.

CHAPTER 5 ANSWERS AND EXPLANATIONS

Multiple-Choice Questions

1. **D**   Na likes to give up an electron (oxidation) in the reaction below.

   $$Na \rightarrow Na^+ + e^-$$

   When a substance likes to be oxidized, it is a strong reducing agent.

2. **B** H^+ is reduced in the reaction below.

   $$2\,H^+ + 2e^- \rightarrow H_2$$

   By definition, the potential for this reaction is zero, so accordingly, the oxidation potential for H_2 is also zero.

3. **A** MnO_4^- likes to gain electrons (reduction) in either of the following reactions:

   $$MnO_4^- + 8\,H^+ + 5e^- \rightarrow Mn_2^+ + 2\,H_2O$$

   $$MnO_4^- + 2\,H_2O + 3e^- \rightarrow MnO_2 + 4\,OH^-$$

   When a substance likes to be reduced, it is a strong oxidizing agent.

4. **C**   The two half-reactions are

   $$2\,Fe^{3+} + 2e^- \rightarrow 2\,Fe^{2+}$$

   $$Mg \rightarrow Mg^{2+} + 2e^-$$

5. **C**   The two half-reactions are

   $$F_2 + 2e^- \rightarrow 2\,F^-$$

   $$2\,Br^- \rightarrow 2e^- + Br_2$$

6. **B**   The two half-reactions are:

   $$Mn^{7+} + 3e^- \rightarrow Mn^{4+}$$
   $$Cr \rightarrow Cr^{3+} = 3e^-$$

7. **D**   The empirical formula of boron trifluoride is BF_3.

   Grams = (moles)(MW)

   Grams of boron = (1 mol)(10.8 g/mol) = 10.8 g

   Grams of fluorine = (3 mol)(19.0 g/mol) = 57.0 g

   So the mass ratio is about 57 to 11, which is about 5.3 to 1.

8. **C** $Moles = \dfrac{grams}{MW}$

$Moles\ of\ CO_2 = \dfrac{(11g)}{(44g/mol)} = \dfrac{1}{4}\,mol$

If $\dfrac{1}{4}$ mole of CO_2 was produced, then $\dfrac{1}{4}$ mole of C was consumed.

Grams = (moles)(MW)

Grams of carbon = $(\dfrac{1}{4}$ mol$)(12$ g/mol$) = 3$ g

So the percent by mass of carbon was $\dfrac{3}{6} = \dfrac{1}{2} = 50\%$

9. **D** $Moles = \dfrac{grams}{MW}$

Moles of calcium hydroxide = $\dfrac{(148\ g)}{(74\ g/mol)} = 2$ moles

Every mole of $Ca(OH)_2$ contains 2 moles of oxygen.

So there are $(2)(2) = 4$ moles of oxygen

Grams = (moles)(MW)

So grams of oxygen = $(4$ mol$)(16$ g/mol$) = 64$ grams

10. **A** Assume that we have 100 grams of the compound. That means that we have 31 grams of oxygen and 69 grams of chlorine.

$Moles = \dfrac{grams}{MW}$

Moles of oxygen = $\dfrac{(31g)}{(16\ g/mol)}$ = slightly less than 2 mol

Moles of chlorine = $\dfrac{(69\ g)}{(35.5\ g/mol)}$ = slightly less than 2 mol

So the ratio of chlorine to oxygen is 1 to 1, and the empirical formula is ClO^-.

11. D $\text{Moles} = \dfrac{\text{liters}}{\left(22.4 \text{ L/mol}\right)}$

$\text{Moles of } CO_2 = \dfrac{67.2 \text{ L}}{\left(22.4 \text{ L/mol}\right)} = 3 \text{ mol}$

According to the balanced equation, if 3 moles of CO_2 were produced, 1 mole of C_3H_8 was consumed.

$\text{Grams} = (\text{moles})(\text{MW})$

So grams of C_3H_8 = (1 mol)(44 g/mol) = 44 grams

12. D The molecular weight of $NaNO_3$ is (23) + (14) + (3)(16) = 85 g/mol.

We can get the answer using pretty rough estimates.

The percent by mass of Na = $\dfrac{(23)}{(85)}$ = between 25% ($\dfrac{1}{4}$) and 33% ($\dfrac{1}{3}$)

The percent by mass of N = $\dfrac{(14)}{(85)}$ = between 10% ($\dfrac{1}{10}$) and 20% ($\dfrac{1}{5}$)

The percent by mass of O = $\dfrac{(48)}{(85)}$ = between 50% ($\dfrac{1}{2}$) and 60% ($\dfrac{3}{5}$)

You can use POE to get choice (D).

13. A The molecular weights of the three compounds are as follows:

Na_2O – 62 g/mol

CaO – 56 g/mol

SiO_2 – 60 g/mol

Because the molecular weights are close together, we can safely say that SiO_2, which makes up a much greater percentage by mass of bottle glass than the other two, will have far and away the most moles in a sample. So the answer must be (A) or (B).

Remember that (grams) = (moles)(MW). Because a sample of bottle glass will have the same number of grams of Na_2O and CaO, the one with the smaller molecular weight must have the greater number of moles. So there must be more moles of CaO than Na_2O.

14.   A   First, you have to remember that 1 liter of water has a mass of 1 kg.

Moles = (molarity)(liters)

Moles of NaCl = (0.5 M)(1 L) = 0.5 moles

Grams = (moles)(MW)

Grams of NaCl = (0.5 mol)(59 g/mol) = 30 g

15.   C   The molecular weight of $CuSO_4$ is 160 g/mol, so we have only 1 mole of the hydrate. The lost mass was due to water, so 1 mole of the hydrate must have contained 90 grams of H_2O.

$$Moles = \frac{grams}{MW}$$

$$Moles\ of\ water = \frac{(90\ g)}{(18\ g/mol)} = 5\ moles$$

So if 1 mole of hydrate contains 5 moles of H_2O, then the formula for the hydrate must be $CuSO_4 \cdot 5\ H_2O$.

16.   B   You might be able to do this one in your head just from knowing that sulfur's molecular weight is twice as large as oxygen's. If not, let's say you have 100 grams of the compound. So you have 50 grams of sulfur and 50 grams of oxygen.

$$Moles = \frac{grams}{MW}$$

$$Moles\ of\ sulfur = \frac{(50\ g)}{(32\ g/mol)} = a\ little\ more\ than\ 1.5$$

$$Moles\ of\ oxygen = \frac{(50\ g)}{(16\ g/mol)} = a\ little\ more\ than\ 3$$

The molar ratio of O to S is 2 to 1, so the empirical formula must be SO_2.

17.   B   Use trial and error or backsolve.

Start with (C). If there are 3 OH^-, there can't be a whole number coefficient for H_2O, so (C) is wrong. You should also be able to see that the answer can't be an odd number, so (A) is are also wrong.

Try (D). If there are 4 OH^-, then there are 2 H_2O.

That leaves 2 more Os on the product side, so there must be 2 CNO^-.

If there are 2 CNO^- then there are 2 CN^-.

These are all whole numbers, but they are not the lowest whole numbers, so (D) is wrong.

If we divide all the coefficients by 2, we get the lowest whole number coefficients. That leaves us with 2 OH^-, which is choice (B).

By the way, N^{5-} (in CN^-) is oxidized to N^{3-} (in CNO^-), so there are 2 e^-.

18. **C** $Moles = \dfrac{grams}{MW}$

Moles of $CaCO_3 = \dfrac{(150\ g)}{(100\ g/mol)} = 1.5$ moles

From the balanced equation, for every mole of $CaCO_3$ consumed, one mole of CO_2 is produced. So 1.5 moles of CO_2 are produced.

At STP, volume of gas = (moles)(22.4 L)

So volume of CO_2 = (1.5)(22.4) = 34 L

19. **B** All of the reactants are consumed in the reaction and the temperature doesn't change, so the pressure will change only if the number of moles of gas changes over the course of the reaction. The number of moles of gas (3 moles) doesn't change in the balanced equation, so the pressure will remain the same (2 atm) at the end of the reaction as at the beginning.

20. **D** Let's say we have 100 grams of the compound.

$Moles = \dfrac{grams}{MW}$

So moles of carbon = $\dfrac{(80\ g)}{(12\ g/mol)} = 6.7$ moles

and moles of hydrogen = $\dfrac{(20\ g)}{(1\ g/mol)} = 20$ moles

According to our rough calculation, there are about three times as many moles of hydrogen in the compound as there are moles of carbon, so the empirical formula is CH_3.

The molar mass for the empirical formula is 15 g/mol, so we need to double the moles of each element to get a compound with a molar mass of 30 g/mol. That makes the molecular formula of the compound C_2H_6.

21.   C   Backsolving doesn't work so well in this case because there are two different compounds that contain oxygen on the right side of the equation, which makes the process kind of confusing. Instead, let's just try plugging in values for the most complicated compound in the equation, $CuFeS_2$.

What if there's 1 $CuFeS_2$? That's impossible because there are 2 Cu's on the right.

What if there are 2 $CuFeS_2$'s? Then the right side has 1 Cu_2S to balance the Cu and 2 FeO to balance the Fe. The right side must also have 3 SO_2 to balance the S.

Now there are 8 O's on the right, so there must be 4 O_2's on the left, and the equation is balanced.

$$2\ CuFeS_2 + 4\ O_2 \rightarrow 1\ Cu_2S + 2\ FeO + 3\ SO_2$$

22.   C   Let's say we have 100 grams of the compound.

$$\text{Moles} = \frac{\text{grams}}{\text{MW}}$$

$$\text{Moles of chlorine} = \frac{(38\ g)}{(35.5\ g/mol)} = \text{about 1 mole}$$

$$\text{Moles of fluorine} = \frac{(62\ g)}{(19\ g/mol)} = \text{about 3 moles}$$

According to our rough calculation, there are about three times as many moles of fluorine in the compound as there are moles of chlorine, so the empirical formula is ClF_3.

23.   C   The oxidation numbers of the reactants are Cr^{6+}, O^{2-}, I^-, and H^+, and the oxidation numbers of the products are Cr^{3+}, O^{2-}, I^0, and H^+.

Chromium gains electrons and is reduced, and iodine loses electrons and is oxidized; the oxidation states of oxygen and hydrogen are not changed.

24. **B** $E°$ for a redox reaction is given by the expression below.

$$E° = E°_{ox} + E°_{red}$$

Al(s) loses 3 electrons in the reaction, so it is oxidized (LEO), and we use the voltage given for the reduction half-reaction, but we change the sign. So $E°_{ox}$ = 1.66 V.

Cr^{3+} gains 3 electrons in the reaction, so it is reduced (GER), and we use the voltage given for the reduction half-reaction. So $E°_{red}$ = –0.74 V.

So, $E°$ = 1.66 V + (–0.74 V) = 0.92 V.

From the relationship, $\Delta G° = -nFE°$, we know that if $E°$ is positive, then $\Delta G°$ is negative, and if $\Delta G°$ is negative, then the reaction is spontaneous under standard conditions.

25. **C** In a molecule, the more electronegative element takes the negative oxidation state. Hydrogen is more electronegative than calcium, so the oxidation state for Ca is +2 and the oxidation state for H is –1.

In choices (A), (B), and (D), hydrogen is the less electronegative element, and it takes the oxidation state of +1.

26. **C** The oxidation state of hydrogen remains +1 throughout the reaction.

We have O$^-$ at the start.

When water forms, oxygen has gained an electron and been reduced to O^{2-} (GER). When O$_2$ forms, oxygen has been oxidized to O^0 through the loss of an electron (LEO). As you can see, in this process, oxygen is both oxidized and reduced.

27. **B** The half-reaction given above is for the oxidation of Fe (LEO). If the presence of Zn impedes this process it is because Zn is oxidized instead of Fe. Iron nails are often coated with zinc to help keep them from rusting.

28. **D** A voltage of 2.0 V is not enough to cause the reduction of Na$^+$ (more than 2.7 V is needed), so no solid sodium will form. A voltage of 2.0 V will cause the reduction of H$_2$O, which forms hydrogen gas, and the oxidation of Cl$^-$, which forms chlorine gas.

From AN OX and RED CAT, we know that hydrogen gas is formed at the cathode and that chlorine gas is formed at the anode.

29.   **A**   We add the reduction potential for Fe^{2+} (–0.4 V) to the oxidation potential for Cu (–0.3 V, the reverse of the reduction potential) to get –0.7 V.

30.   **D**   To get the reaction potential, we add the reduction potential for the reduction half-reaction to the oxidation potential for the oxidation half–reaction.

Here are the reaction potentials for choices (A)–(D).

(A) (–1.2 V) + (–0.3 V) = –1.5 V

(B) (–1.2 V) + (+0.8 V) = –0.4 V

(C) (–0.8 V) + (–0.3 V) = –1.1 V

(D) (–0.8 V) + (+1.2 V) = +0.4 V

Choice (D) is the only reaction with a positive voltage, which means that it is the only reaction listed that occurs spontaneously.

31.   **D**   In order for the metallic species to be oxidized the reduction potential of the oxidant (F_2 or Cl_2) must be greater than the reduction potential of the metallic species. Cl_2 is a strong enough oxidant to take Ag° to Ag^+, but not strong enough to take Ag^+ to Ag^{2+} or Au° to Au^{3+}. On the other hand, F_2 *is* a strong enough oxidant, as it has the highest reduction potential listed, to do all of the above electrochemical reactions. Therefore Au° will be oxidized to Au^{3+} and Ag° will be oxidized all the way to Ag^{2+}.

32.   **B**   The spontaneity of the disproportionation reaction means that the standard potentials of the following two half reactions must sum to a positive number:

$M^+ + 1e^- \rightarrow M^\circ$

$M^+ \rightarrow 1e^- + M^{2+}$

The two potentials here are the reduction potential and oxidation potential of M^+. Without knowing the relative values and signs of the two individual reactions we cannot say for sure any of the rest of the listed pairs of potentials would sum to a positive.

33. **C** First let's find out how many electrons are provided by the current.

Coulombs = (seconds)(amperes) = (600)(5.00)

$$\text{Moles of electrons} = \frac{(\text{coulombs})}{(96,500)} = \frac{(600)(5.00)}{(96,500)} \text{ moles}$$

In $AlCl_3$, we have Al^{3+}, so the half-reaction for the plating of aluminum is

$$Al^{3+} + 3e^- \rightarrow Al(s)$$

So we get $\frac{1}{3}$ as many moles of $Al(s)$ as we have moles of electrons.

$$\text{Moles of Al}(s) = (\text{moles of electrons})(\frac{1}{3}) = \frac{(600)(5.00)}{(96,500)(3)} \text{ moles}$$

Now, grams = (moles)(MW)

$$\text{Grams of Al}(s) = \frac{(600)(5.00)}{(96,500)(3)}(27.0) = \frac{(600)(5.00)(27.0)}{(96,500)(3)} \text{ grams}$$

Free Response Questions

1. (a) $CaCO_3 + 2\,HCl \rightarrow CaCl_2 + H_2O + CO_2$

    (b) Use the ideal gas equation to find the number of moles of CO_2 produced. Remember to convert to the proper units.

    $$n = \frac{PV}{RT} = \frac{\left(\frac{740}{760}\text{ atm}\right)(0.900\text{ L})}{(0.0821)(303\text{ K})} = 0.035 \text{ moles}$$

    From the balanced equation, for every mole of CO_2 produced, 1 mole of $CaCO_3$ was consumed. So 0.035 moles of $CaCO_3$ were consumed.

    (c) We know the number of moles of $CaCO_3$, so we can find the mass.

    Grams = (moles)(MW)

    Grams of $CaCO_3$ = (0.035 mol)(100 g/mol) = 3.50 grams

    $$\text{Percent by mass} = \frac{\text{mass of CaCO}_3}{\text{mass of sample}} \times 100 = \frac{(3.50\text{ g})}{(10.0\text{ g})} \times 100 = 35\%$$

    (d) Mass of SiO_2 = 10.0 g – 3.5 g = 6.5 g

    $$\text{Moles} = \frac{\text{grams}}{\text{MW}}$$

    $$\text{Moles of SiO}_2 = \frac{(6.5\text{ g})}{(60\text{ g/mol})} = 0.11 \text{ mol}$$

    $$\text{Molar ratio} = \frac{\text{moles of CaCO}_3}{\text{moles of SiO}_2} = \frac{(0.035)}{(0.11)} = 0.32$$

2. (a) All of the hydrogen in the water and all of the carbon in the carbon dioxide must have come from the hydrocarbon.

$$\text{Moles of } H_2O = \frac{(2.16 \text{ g})}{(18.0 \text{ g/mol})} = 0.120 \text{ moles}$$

Every mole of water contains 2 moles of hydrogen, so there are 0.240 moles of hydrogen.

$$\text{Moles of } CO_2 = \frac{(1.80 \text{ L})}{(22.4 \text{ L/mol})} = 0.080 \text{ moles}$$

Every mole of CO_2 contains 1 mole of carbon, so there are 0.080 moles of carbon.

There are three times as many moles of hydrogen as there are moles of carbon, so the empirical formula of the hydrocarbon is CH_3.

(b) In (a), we found the number of moles of hydrogen and carbon consumed, so we can find the mass of the hydrocarbon.

Grams = (moles)(MW)

Grams of H = (0.240 mol)(1.01 g/mol) = 0.242 g

Grams of C = (0.080 mol)(12.01 g/mol) = 0.961 g

Grams of hydrocarbon = (0.242) + (0.961) = 1.203 g

(c) First let's find the number of moles of hydrocarbon from the ideal gas law. Don't forget to convert to the appropriate units (760 mmHg = 1 atm, 32°C = 305 K).

$$n = \frac{PV}{RT} = \frac{(1.00 \text{ atm})(1.00 \text{ L})}{(0.0821)(305 \text{ K})} = 0.040 \text{ moles}$$

Now we can use the mass we found in (b) to find the molecular weight of the hydrocarbon.

$$MW = \frac{\text{grams}}{\text{moles}} = \frac{(1.203 \text{ g})}{(0.040 \text{ mol})} = 30.1 \text{ g/mole}$$

CH_3 would have a molecular weight of 15, so we can just double the empirical formula to get the molecular formula, which is C_2H_6.

(d) $2 C_2H_6 + 7 O_2 \rightarrow 4 CO_2 + 6 H_2O$

3. (a) First find the molecular weight of Cu_2S.

MW of Cu_2S = (2)(63.6) + (1)(32.1) = 159.3% by mass of

$$Cu = \frac{\text{mass of Cu}}{\text{mass of Cu}_2S} \times 100$$

$$= \frac{(2)(63.6)}{(159.3)} \times 100 = 79.8\%$$

(b) Assume that we have 100 grams of ore #2. So we have 34.6 g of Cu, 30.5 g of Fe, and 34.9 g of S. To get the empirical formula, we need to find the number of moles of each element.

$$\text{Moles} = \frac{\text{grams}}{\text{MW}}$$

$$\text{Moles of Cu} = \frac{(34.6 \text{ g})}{(63.6 \text{ g/mol})} = 0.544 \text{ moles of Cu}$$

$$\text{Moles of Fe} = \frac{(30.5 \text{ g})}{(55.9 \text{ g/mol})} = 0.546 \text{ moles of Fe}$$

$$\text{Moles of S} = \frac{(34.9 \text{ g})}{(32.1 \text{ g/mol})} = 1.09 \text{ moles of S}$$

So the molar ratio of Cu:Fe:S is 1:1:2, and the empirical formula for

ore #2 is $CuFeS_2$.

(c) You can use the ratio of the percents by weight.

$$\text{Mass of S} = \frac{\% \text{ by mass of S}}{\% \text{ by mass of Fe}} \times (\text{mass of Fe}) = \frac{(28.1 \%)}{(16.3 \%)} (11.0 \text{ g}) = 19.0 \text{ g}$$

(d) First find the moles of O_2 consumed.

$$\text{Moles} = \frac{\text{grams}}{\text{MW}}$$

$$\text{Moles of O}_2 = \frac{(3.84 \text{ g})}{(32.0 \text{ g/mol})} = 0.120 \text{ moles}$$

From the balanced equation, for every 3 moles of O_2 consumed, 6 moles of Cu are produced, so the number of moles of Cu produced will be twice the number of moles of O_2 consumed. So 0.240 moles of Cu are produced.

Grams = (moles)(MW)

Grams of Cu = (0.240 mol)(63.6 g/mol) = 15.3 grams

4. (a) We need to find the limiting reagent. There's plenty of water, so it must be one of the other two reactants.

$$\text{Moles} = \frac{\text{grams}}{\text{MW}}$$

$$\text{Moles of Mg} = \frac{(1.46 \text{ g})}{(24.3 \text{ g/mol})} = 0.060 \text{ moles}$$

$$\text{Moles} = (\text{molarity})(\text{volume})$$

$$\text{Moles of CuSO}_4 = (0.200 \text{ M})(0.500 \text{ L}) = 0.100 \text{ moles}$$

From the balanced equation, Mg and $CuSO_4$ are consumed in a 1:1 ratio, so we'll run out of Mg first. Mg is the limiting reagent, and we'll use it to find the yield of H_2.

From the balanced equation, 1 mole of H_2 is produced for every 2 moles of Mg consumed, so the number of moles of H_2 produced will be half the number of moles of Mg consumed.

$$\text{Moles of H}_2 = \frac{1}{2}(0.060 \text{ mol}) = 0.030 \text{ moles}$$

(b) Mg is the limiting reagent, so some $CuSO_4$ will remain. From the balanced equation, Mg and $CuSO_4$ are consumed in a 1:1 ratio, so when 0.060 moles of Mg are consumed, 0.060 moles of $CuSO_4$ are also consumed.

$$\text{Moles of CuSO}_4 \text{ remaining} = (0.100 \text{ mol}) - (0.060 \text{ mol}) = 0.040 \text{ moles}$$

(c) From the balanced equation, 1 mole of Cu_2O is produced for every 2 moles of Mg consumed, so the number of moles of Cu_2O produced will be half the number of moles of Mg consumed.

$$\text{Moles of Cu}_2\text{O} = \frac{1}{2}(0.060 \text{ mol}) = 0.030 \text{ moles}$$

$$\text{Grams} = (\text{moles})(\text{MW})$$

$$\text{Grams of Cu}_2\text{O} = (0.030 \text{ mol})(143 \text{ g/mol}) = 4.29 \text{ grams}$$

(d) All of the Mg consumed ends up as Mg^{2+} ions in the solution.

$$\text{Molarity} = \frac{\text{moles}}{\text{liters}}$$

$$[Mg^{2+}] = \frac{(0.060 \text{ mol})}{(0.500 \text{ L})} = 0.120 \text{ M}$$

5. (a) Silver half-reaction: $Ag^+ + e^- \rightarrow Ag(s)$

Silver gains an electron, so it is reduced (GER). Reduction takes place at the cathode (RED CAT).

Water half-reaction: $2\,H_2O \rightarrow O_2 + 4\,H^+ + 4e^-$

Water gives up electrons, so it is oxidized (LEO). Oxidation takes place at the anode (AN OX).

(b) First let's find out how many electrons were supplied by the current.

Coulombs = (amperes)(seconds)

Coulombs = (2.00 A)(10.0 min)(60 sec/min) = 1,200 C

$$\text{Moles of electrons} = \frac{\text{coulombs}}{96,500} = \frac{1,200}{96,500} = 0.0124 \text{ moles}$$

From the water half-reaction we know that one mole of oxygen gas is produced for every 4 electrons given up, so there will be $\frac{1}{4}$ as many moles of O_2 as there are electrons.

Moles of O_2 = $(\frac{1}{4})$(moles of electrons) = $(\frac{1}{4})$(0.0124 mol) = 3.10 × 10^{-3} moles

At STP, volume of gas = (moles)(22.4 L/mol)

Volume of O_2 = (3.10 × 10^{-3} mol)(22.4 L/mol) = 0.0694 L

(c) From (b), we know that the current provided 0.0124 moles of electrons.

From the silver half-reaction, we know that for every mole of electrons consumed, one mole of Ag(s) is produced. So we have 0.0124 moles of Ag(s).

Grams = (moles)(MW)

Grams of Ag(s) = (0.0124 mol)(107.87 g/mol) = 1.34 grams

(d) From (b), we know that the current provided 0.0124 moles of electrons. From the water half-reaction we know that 4 moles of H^+ were produced for every 4 electrons given up, so the number of moles of H^+ will be the same as the number of electrons.

Moles of H^+ = 0.0124

We can assume that the volume of solution doesn't change during the electrolysis.

$$\text{Molarity} = \frac{\text{moles}}{\text{volume}}$$

$$[H^+] = \frac{(0.0124 \text{ mol})}{(0.800 \text{ L})} = 0.0155 \; M$$

$$pH = -\log[H^+] = -\log(0.0155) = 1.81$$

6. (a) The salt bridge is necessary to maintain electrical neutrality in the two compartments where the half-reactions are taking place. It does this by allowing anions (–) to move into the anode compartment and cations (+) to move into the cathode compartment.

(b) From the table of standard reduction potentials, we can see that Fe wants to be oxidized (reading the table from right to left gives a positive oxidation potential), so it can act as the reducing agent for H^+ ions.

Cu does not want to be oxidized (reading the table [from right to left] gives a negative oxidation potential), so it will not reduce the H^+ ions.

(c) (i) Use the table of standard reduction potentials.

$$Ag^+ + e^- \rightarrow Ag(s) \qquad E° = +0.80 \text{ V}$$

$$Sn(s) \rightarrow Sn^{2+} + e^- \qquad E° = \underline{+0.14 \text{ V}}$$

$$+0.94 \text{ V}$$

(ii) Use the Nernst equation.

$$E_{cell} = E° - \frac{0.0592}{n} \log Q$$

$$E_{cell} = E° - \frac{0.0592}{n} \log \frac{\left[Sn^{2+}\right]}{\left[Ag^+\right]}$$

When $[Ag^+]$ is increased, $\log \dfrac{\left[Sn^{2+}\right]}{\left[Ag^+\right]}$ decreases, which means that a smaller number is subtracted from $E°$, which means that E_{cell} increases.

You can also use Le Châtelier's law. Increasing $[Ag^+]$ increases the concentration of the reactants, which favors the forward reaction and increases the voltage of the cell.

(iii) The cell potential changes with changing concentrations. The concentration of a solid is constant, so changing the amount of solid Sn present in the cell will have no effect on the cell potential.

7. (a) The reduction potential for Cd^{2+} is equal to -0.40 V. For the reaction to proceed spontaneously, the cell potential must be greater than zero, so the oxidation potential for M(s) must be greater than $+0.40$ V. That means that the reduction potential must be less than -0.40V.

(b) $M(s) \rightarrow M^+ + e^-$

M(s) loses an electron and is oxidized (LEO), so this half-reaction takes place at the anode (AN OX).

$Cd^{2+} + 2e^- \rightarrow Cd(s)$

Cd^{2+} gains electrons and is reduced (GER), so this half-reaction takes place at the cathode (RED CAT).

(c) (i) As the reaction proceeds in the forward direction, more M^+ will be generated, so $[M^+]$ will increase.

(ii) Use the Nernst equation.

$$E_{cell} = E° - \frac{0.0592}{n} \log Q$$
$$E_{cell} = E° - \frac{0.0592}{n} \log \frac{[M^+]}{[Cd^{2+}]}$$

As the reaction proceeds, $[M^+]$ increases and $[Cd^{2+}]$ decreases, thereby increasing the log value, which means that progressively larger numbers are subtracted from $E°$, which means that E_{cell} decreases.

(iii) Use the equation

$\Delta G = -nFE$

From (ii), we know that E is decreasing. As E decreases, G becomes less negative, so it increases. That means that as the reaction proceeds, it becomes less spontaneous.

8. (a) $Na^+ + e^- \rightarrow Na$

        Na^+ gains an electron, so it is reduced (GER).

        Reduction takes place at the cathode (RED CAT).

        $2\ Cl^- \rightarrow Cl_2 + 2e^-$

        Cl^- loses an electron, so it is oxidized (LEO).

        Oxidation takes place at the anode (AN OX).

    (b)   Chlorine gas (Cl_2) bubbles out of the molten mixture, leaving pure molten sodium.

    (c)   (i) $E_{reaction}$ will be negative because E for both half-reactions is negative. The voltage source that drives the reaction must be strong enough to overcome the negative voltage for the reaction.

          (ii) ΔG will be positive because the reaction is not spontaneous. Also, from the relationship $\Delta G = -nFE$ we know that ΔG and E always have opposite signs.

          (iii) ΔS will be positive because the reactant is a liquid and the products are a liquid and a gas.

    (d)   Water molecules are more easily reduced than sodium ions, so the hydrogen in H_2O will be reduced at the cathode instead of Na^+.

Chapter 6
Big Idea #4:
Chemical Reactions
and their Rates

Rates of chemical reactions are determined by
details of the molecular collisions.

RATE LAW USING INITIAL CONCENTRATIONS

The rate law for a reaction describes the dependence of the initial rate of a reaction on the concentrations of its reactants. It includes the Arrhenius constant, k, which takes into account the activation energy for the reaction and the temperature at which the reaction occurs. The rate of a reaction is described in terms of the rate of appearance of a product or the rate of disappearance of a reactant. The rate law for a reaction cannot be determined from a balanced equation; it must be determined from experimental data, which is presented on the test in table form.

Here's how it's done

The data below were collected for the following hypothetical reaction:

$$A + 2B + C \rightarrow D$$

Experiment	Initial Concentration of Reactants (M)			Initial Rate of Formation of D (M/sec)
	[A]	[B]	[C]	
1	0.10	0.10	0.10	0.01
2	0.10	0.10	0.20	0.01
3	0.10	0.20	0.10	0.02
4	0.20	0.20	0.10	0.08

The rate law always takes the following form, using the concentrations of the reactants:

$$\text{Rate} = k[A]^x[B]^y[C]^z$$

The greater the value of a reactant's exponent, the more a change in the concentration of that reactant will affect the rate of the reaction. To find the values for the exponents x, y, and z, we need to examine how changes in the individual reactants affect the rate. The easiest way to find the exponents is to see what happens to the rate when the concentration of an individual reactant is doubled.

Let's look at [A]

From experiment 3 to experiment 4, [A] doubles while the other reactant concentrations remain constant. For this reason, it is useful to use the rate values from these two experiments to calculate x (the order of the reaction with respect to reactant A).

As you can see from the table, the rate quadruples from experiment 3 to experiment 4, going from 0.02 M/sec to 0.08 M/sec.

We need to find a value for the exponent x that relates the doubling of the concentration to the quadrupling of the rate. The value of x can be calculated as follows:

$$(2)^x = 4, \text{ so } x = 2$$

Because the value of x is 2, the reaction is said to be second order with respect to A.

$$Rate = k[A]^2[B]^y[C]^z$$

Let's look at [B]

From experiment 1 to experiment 3, [B] doubles while the other reactant concentrations remain constant. For this reason it is useful to use the rate values from these two experiments to calculate y (the order of the reaction with respect to reactant B).

As you can see from the table, the rate doubles from experiment 1 to experiment 3, going from 0.01 M/sec to 0.02 M/sec.

We need to find a value for the exponent y that relates the doubling of the concentration to the doubling of the rate. The value of y can be calculated as follows:

$$(2)^y = 2, \text{ so } y = 1$$

Because the value of y is 1, the reaction is said to be first order with respect to B.

$$Rate = k[A]^2[B][C]^z$$

Let's look at [C]

From experiment 1 to experiment 2, [C] doubles while the other reactant concentrations remain constant.

The rate remains the same at 0.01 M.

The rate change is $(2)^z = 1$, so z = 0.

Because the value of z is 0, the reaction is said to be zero order with respect to C.

$$Rate = k[A]^2[B]$$

Because the sum of the exponents is 3, the reaction is said to be third order overall.

Once the rate law has been determined, the value of the rate constant can be calculated using any of the lines of data on the table. The units of the rate constant are dependent on the order of the reaction, so it's important to carry along units throughout all rate constant calculations.

Let's use experiment 3.

$$k = \frac{Rate}{[A]^2[B]} = \frac{(0.02\,M\,/\,\sec)}{(0.10\,M)^2(0.20\,M)} = 10\left(\frac{(M)}{(M)^3(\sec)}\right) = 10\,M^{-2}\text{-}\sec^{-1}$$

You should note that we can tell from the coefficients in the original balanced equation that the rate of appearance of D is equal to the rate of disappearance of A and C because the coefficients of all three are the same. The coefficient of D is half as large as the coefficient of B, however, so the rate at which D appears is half the rate at which B disappears.

RATE LAW USING CONCENTRATION AND TIME

It's also useful to have rate laws that relate the rate constant k to the way that concentrations change over time. The rate laws will be different depending on whether the reaction is first, second, or zero order, but each rate law can be expressed as a graph that relates the rate constant, the concentration of a reactant, to the elapsed time.

First-Order Rate Laws

The rate of a first-order reaction depends on the concentration of a single reactant raised to the first power.

$$\text{Rate} = k[A]$$

As the concentration of reactant A is depleted over time, the rate of reaction will decrease with a characteristic half-life. This is the same curve you've seen used for nuclear decay, which is also a first-order process.

The rate law for a first-order reaction uses natural logarithms.

First-Order Rate Law

$$\ln[A]_t - \ln[A]_o = -kt$$

$[A]_t =$ concentration of reactant A at time t

$[A]_o =$ initial concentration of reactant A

$k =$ rate constant

$t =$ time elapsed

The use of natural logarithms in the rate law creates a linear graph comparing concentration and time. The slope of the line is given by $-k$ and the y-intercept is given by $\ln[A]_o$.

Time

If you are asked to give the half-life of a first-order reaction, you can usually figure it out by looking at the graph or table given in the problem, but it's worth your while to know the equation that gives the half-life in terms of the rate constant.

$$\text{Half-life} = \frac{\ln 2}{k} = \frac{0.693}{k}$$

Let's try an example based on the data below.

[A] (M)	Time (min)
2.0	0
1.6	5
1.2	10

(a)   Let's find the value of k. We'll use the first two lines of the table.
$$\ln[A]_t - \ln[A]_o = -kt$$
$$\ln(1.6) - \ln(2.0) = -k(5 \text{ min})$$
$$-0.22 = -(5 \text{ min})k$$
$$k = 0.045 \text{ min}^{-1}$$

(b)   Now let's use k to find [A] when 20 minutes have elapsed.
$$\ln[A]_t - \ln[A]_o = -kt$$
$$\ln[A]_t - \ln(2.0) = -(0.045 \text{ min}^{-1})(20 \text{ min})$$
$$\ln[A]_t = -0.21$$
$$[A]_t = e^{-0.21} = 0.81 \ M$$

(c)   Now let's find the half-life of the reaction.

We can look at the answer (b) to see that the concentration dropped by half (1.6 M to 0.8 M) from 5 minutes to 20 minutes. That makes the half-life about 15 minutes. We can confirm this using the half-life equation.

$$\text{Half-life} = \frac{0.693}{k} = \frac{0.693}{0.045 \text{ min}^{-1}} = 15.4 \text{ minutes}$$

Second-Order Rate Laws

The rate of a second-order reaction depends on the concentration of a single reactant raised to the second power.

$$\text{Rate} = k[A]^2$$

The rate law for a second-order reaction uses the inverses of the concentrations.

Second-Order Rate Law

$$\frac{1}{[A]_t} - \frac{1}{[A]_o} = kt$$

$[A]_t$ = concentration of reactant A at time t

$[A]_o$ = initial concentration of reactant A

k = rate constant

t = time elapsed

The use of inverses in the rate law creates a linear graph comparing concentration and time. Notice that the line moves upward as the concentration decreases. The slope of the line is given by k and the y-intercept is given by $\dfrac{1}{[A]_o}$.

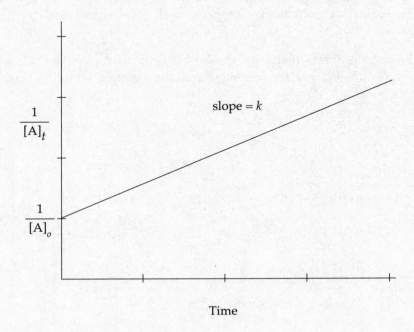

Zero-Order Rate Laws

The rate of a zero-order reaction does not depend on the concentration of reactants at all, so the rate of a zero-order reaction will always be the same at a given temperature.

$$\text{Rate} = k$$

The graph of the change in concentration of a reactant of a zero-order reaction versus time will be a straight line with a slope equal to $-k$.

COLLISION THEORY

According to collision theory, chemical reactions occur because reactants are constantly moving around and colliding with one another.

When reactants collide with sufficient energy (**activation energy, E_a**), a reaction occurs. These collisions are referred to as effective collisions, because they lead to a chemical reaction. Ineffective collisions do not produce a chemical reaction. At any given time during a reaction, a certain fraction of the reactant molecules will collide with sufficient energy to cause a reaction between them.

Reaction rate increases with increasing concentration of reactants because if there are more reactant molecules moving around in a given volume, then more collisions will occur.

Reaction rate increases with increasing temperature because increasing temperature means that the molecules are moving faster, which means that the molecules have greater average kinetic energy. The higher the temperature, the greater the number of reactant molecules colliding with each other with enough energy (E_a) to cause a reaction.

The Boltzmann distribution diagram below is often used to show that increasing temperature increases the fraction of reactant molecules above the activation energy.

The rate constant will increase when temperature is increased. The relationship between temperature and the rate constant is given by the Arrhenius equation, $k = Ae^{\frac{-E_a}{RT}}$. This equation is generally rewritten using natural logarithms.

The Arrhenius Equation

$$\ln k = -\frac{E_a}{R}\left(\frac{1}{T}\right) + \ln A$$

k = rate constant
E_a = activation energy
R = gas constant, 8.31 J/k-mol
T = absolute temperature (K)
A = a constant that takes into account collision frequency and orientation

When $\ln k$ is graphed versus $\dfrac{1}{T}$, it makes a straight line with a slope of $-\dfrac{E_a}{R}$. It's a little difficult to see, but the graph below shows that k is increasing as the temperature increases.

REACTION MECHANISMS

Many chemical reactions are not one-step processes. Rather, the balanced equation is the sum of a series of simple steps. Three molecules will not collide simultaneously very often, so steps of a reaction mechanism involve only one or two reactants at a time.

For instance, the hypothetical reaction

$$2\,A + 2\,B \rightarrow C + D \qquad\qquad \text{Rate} = k[A]^2[B]$$

could take place by the following three-step mechanism:

    I. $A + A \rightleftharpoons X$ (fast)

    II. $X + B \rightarrow C + Y$ (slow)

    III. $Y + B \rightarrow D$ (fast)

Species X and Y are called **intermediates** because they appear in the mechanism, but they cancel out of the balanced equation. The steps of a reaction mechanism must add up to equal the balanced equation, with all intermediates cancelling out.

Let's show that the mechanism above is consistent with the balanced equation by adding up all the steps.

I. $A + A \rightleftharpoons X$

II. $X + B \rightarrow C + Y$

III. $Y + B \rightarrow D$

_____

$A + A + X + B + Y + B \rightarrow X + C + Y + D$

Cancel species that appear on both sides.

$2\,A + 2\,B \rightarrow C + D$

By adding up all the steps, we get the balanced equation for the overall reaction, so this mechanism is consistent with the balanced equation.

As in any process where many steps are involved, the speed of the whole process can't be faster than the speed of the slowest step in the process, so the slowest step of a reaction is called the **rate-determining step**. Because the slowest step is the most important step in determining the rate of a reaction, the slowest step and the steps leading up to it are used to see if the mechanism is consistent with the rate law for the overall reaction.

Let's look at the reaction and the three-step mechanism again.

$2\,A + 2\,B \rightarrow C + D$ $\qquad\qquad\qquad$ Rate = $k[A]^2[B]$

The reaction above takes place by the following three-step mechanism.

I. $A + A \rightleftharpoons X$ $\qquad$ (fast)

II. $X + B \rightarrow C + Y$ $\qquad$ (slow)

III. $Y + B \rightarrow D$ $\qquad$ (fast)

Let's show that the reaction mechanism is consistent with the rate law (Rate = $k[A]^2[B]$).

The slowest step is the rate-determining step, so we should start with the rate law for step II:

$$\text{Rate} = k_2[X][B]$$

But X is an intermediate, which means that it can't appear in the overall rate law. To eliminate X from the rate law, we need to look at the equilibrium reaction in step I. We can assume that the reaction in step I comes to equilibrium quickly. At equilibrium the rate of the forward reaction is equal to the rate of the reverse reaction, so we get

$$k_f[A][A] = k_r[X]$$

Now we can solve for [X].

$$[X] = \frac{k_f}{k_r}[A]^2$$

Once we have solved our equilibrium rate expression for [X] in terms of [A], we can substitute for [X] in our step II rate law.

$$\text{Rate} = k_2\frac{k_f}{k_r}[A]^2[B] = k[A]^2[B]$$

Now we have a rate law that contains only reactants from the overall equation and that's consistent with the experimentally derived rate law that we were given. You can always eliminate intermediates from the rate-determining step by this process.

CATALYSTS

As we mentioned earlier in the book, a catalyst increases the rate of a chemical reaction without being consumed in the process; catalysts do not appear in the balanced equation. In some cases, a catalyst is a necessary part of a reaction because in its absence, the reaction would proceed at too slow a rate to be at all useful.

A catalyst increases the rate of a chemical reaction by providing an alternative reaction pathway with a lower activation energy.

Kinetics and Equilibrium

There is a relationship between the rate constants for the forward and reverse directions of a particular reaction and the equilibrium constant for that reaction.

$$K_{eq} = \frac{k_f}{k_r}$$

K_{eq} = the equilibrium constant
k_f = the rate constant for the forward reaction
k_r = the rate constant for the reverse reaction

RADIOACTIVE DECAY AND HALF-LIFE

Alpha Emission ($_2^4\alpha$)

In **alpha decay**, the nucleus emits a particle that has the same constitution as a helium molecule, with 2 protons and 2 neutrons. Alpha particles are the least penetrating of the products of nuclear decay.

$$_2^4\alpha = {}_2^4\text{He}$$

When a nucleus undergoes alpha decay

- subtract 4 from the **mass number**
- subtract 2 from the atomic number

$$_{92}^{238}\text{U} \rightarrow {}_2^4\alpha + {}_{90}^{234}\text{Th}$$

Beta Emission ($_{-1}^{0}\beta$)

A beta particle is identical to an electron. In beta decay, the nucleus changes a neutron into a proton and an electron and emits the electron.

$$_0^1\text{n} \rightarrow {}_{-1}^0\beta + {}_1^1\text{P}$$

When a nucleus undergoes beta decay

- the mass number remains the same
- add 1 to the atomic number

$$_6^{14}\text{C} \rightarrow {}_{-1}^0\beta + {}_7^{14}\text{N}$$

Positron Emission ($_{+1}^{0}\beta$)

A positron is like an electron with a positive charge. In positron emission, the nucleus changes a proton into a neutron and a positron and emits the positron.

$$_1^1\text{P} \rightarrow {}_{+1}^0\beta + {}_0^1\text{n}$$

When a nucleus undergoes positron emission

- the mass number remains the same
- subtract 1 from the atomic number

$$_5^8\text{B} \rightarrow {}_{+1}^0\beta + {}_4^8\text{Be}$$

Electron Capture ($_{-1}^{0}e$)

In electron capture, the nucleus captures a low energy electron and combines it with a proton to form a neutron.

$$_{-1}^{0}e + {}_{1}^{1}p \rightarrow {}_{0}^{1}n$$

When a nucleus undergoes electron capture

- the mass number remains the same
- subtract 1 from the atomic number

$$_{9}^{18}F + {}_{-1}^{0}e \rightarrow {}_{8}^{18}O$$

Gamma Rays ($_{0}^{0}\gamma$)

Gamma rays are electromagnetic radiation and have no mass and no charge. Gamma rays usually accompany other forms of nuclear decay, and are the most penetrating of the nuclear decay products.

Nuclear Stability

Nuclei undergo decay to achieve greater stability. You can use the periodic table to predict the kind of decay that an isotope will undergo.

- If an isotope's *mass number is greater than its atomic weight* ($_{6}^{16}C$, for instance), the nucleus will try to gain protons and lose neutrons; therefore, if its mass number is greater than its atomic weight, you can *expect beta decay.*

- If an isotope's *mass number is less than its atomic weight* ($_{6}^{11}C$, for instance), the nucleus will try to lose protons and gain neutrons; therefore, if its mass number is less than its atomic weight, you can *expect positron emission* or *electron capture.*
- Alpha emission is seen mainly in very large nuclei, usually with atomic numbers of 60 or greater.

Half-Life

The half-life of a radioactive substance is the time it takes for half of the substance to decay. Most half-life problems can be solved by using a simple chart.

Time	Sample
0	100%
1 half-life	50%
2 half-lives	25%
3 half-lives	12.5%

(Don't forget that the chart should start with the time at zero.)

So a sample with a mass of 120 grams and a half-life of 3 years will decay as follows:

Time	Sample
0	120 g
3	60 g
6	30 g
9	15 g

The fact that you're not allowed to use a calculator for Section I and that you're not given any half-life formulas for Section II means that you should be able to solve any half-life problem that comes up by using the chart and POE.

Mass Defect and Binding Energy

When protons and neutrons come together to form a nucleus, the mass of the nucleus is less than the sum of the masses of its constituent protons and neutrons. This difference in mass is called the **mass defect**. The mass lost in this process is released in the form of energy. If we reverse the process this is the same amount of energy, called the **binding energy**, required to decompose the nucleus back into protons and neutrons. The relationship between mass and energy is given by Einstein's famous equation.

$$E = mc^2$$

E = energy (J)
m = mass (kg)
c = the speed of light, 3×10^8 m/sec

You can see that because c^2 is such a large number, a very small change in mass results in a very large change in energy.

CHAPTER 6 QUESTIONS

Multiple-Choice Questions

Questions 1–3

$$A + B \rightarrow C$$

The following are possible rate laws for the hypothetical reaction given above.

(A) Rate = $k[A]$
(B) Rate = $k[A][B]$
(C) Rate = $k[A]^2[B]$
(D) Rate = $k[A]^2[B]^2$

1. This is the rate law for a first **order reaction.**

2. This is the rate law for a reaction that is second order with respect to B.

3. This is the rate law for a third order reaction.

Questions 4–6

$$A + B \rightarrow C$$

The following are possible rate laws for the hypothetical reaction given above.

(A) Rate = $k[A]$
(B) Rate = $k[B]^2$
(C) Rate = $k[A][B]$
(D) Rate = $k[A]^2[B]$

4. When [A] and [B] are doubled, the initial rate of reaction will increase by a factor of eight.

5. When [A] and [B] are doubled, the initial rate of reaction will increase by a factor of two.

6. When [A] is doubled and [B] is held constant, the initial rate of reaction will not change.

7. A multistep reaction takes place by the following mechanism:

$$A + B \rightarrow C + D$$
$$A + C \rightarrow D + E$$

Which of the species shown above is an intermediate in the reaction?

(A) A
(B) B
(C) C
(D) D

8.
$$2 \, NOCl \rightarrow 2 \, NO + Cl_2$$

The reaction above takes place with all of the reactants and products in the gaseous phase. Which of the following is true of the relative rates of disappearance of the reactants and appearance of the products?

(A) NO appears at twice the rate that NOCl disappears.
(B) NO appears at the same rate that NOCl disappears.
(C) NO appears at half the rate that NOCl disappears.
(D) Cl_2 appears at the same rate that NOCl disappears.

9.
$$H_2(g) + I_2(g) \rightarrow 2 \, HI(g)$$

When the reaction given above takes place in a sealed isothermal container, the rate law is

$$Rate = k[H_2][I_2]$$

If a mole of H_2 gas is added to the reaction chamber, which of the following will be true?

(A) The rate of reaction and the rate constant will increase.
(B) The rate of reaction and the rate constant will not change.
(C) The rate of reaction will increase and the rate constant will decrease.
(D) The rate of reaction will increase and the rate constant will not change.

10.
$$A + B \rightarrow C$$

When the reaction given above takes place, the rate law is Rate = $k[A]$

If the temperature of the reaction chamber were increased, which of the following would be true?

(A) The rate of reaction and the rate constant will increase.
(B) The rate of reaction and the rate constant will not change.
(C) The rate of reaction will increase and the rate constant will decrease.
(D) The rate of reaction will increase and the rate constant will not change.

11.

$$A + B \rightarrow C$$

Based on the following experimental data, what is the rate law for the hypothetical reaction given above?

Experiment	[A] (M)	[B] (M)	Initial Rate of Formation of C (mol/L-sec)
1	0.20	0.10	3×10^{-2}
2	0.20	0.20	6×10^{-2}
3	0.40	0.20	6×10^{-2}

(A) Rate = $k[A]$
(B) Rate = $k[A]^2$
(C) Rate = $k[B]$
(D) Rate = $k[A][B]$

12.

$$A + B \rightarrow C + D$$

The rate law for the hypothetical reaction shown above is as follows:

$$Rate = k[A]$$

Which of the following changes to the system will increase the rate of the reaction?

I. An increase in the concentration of A
II. An increase in the concentration of B
III. An increase in the temperature

(A) I only
(B) I and II only
(C) I and III only
(D) II and III only

13.

$$A + B \rightarrow C$$

Based on the following experimental data, what is the rate law for the hypothetical reaction given above?

Experiment	[A] (M)	[B] (M)	Initial Rate of Formation of C (M/sec)
1	0.20	0.10	2.0×10^{-6}
2	0.20	0.20	4.0×10^{-6}
3	0.40	0.40	1.6×10^{-5}

(A) Rate = $k[A]$
(B) Rate = $k[A]^2$
(C) Rate = $k[B]$
(D) Rate = $k[A][B]$

14.

Time (Hours)	[A] M
0	0.40
1	0.20
2	0.10
3	0.05

Reactant A underwent a decomposition reaction. The concentration of A was measured periodically and recorded in the chart above. Based on the data in the chart, which of the following is the rate law for the reaction?

(A) Rate = k[A]
(B) Rate = k[A]^2
(C) Rate = 2k[A]
(D) Rate = $\frac{1}{2}k$[A]

15. Strontium-90 decays through the emission of beta particles. It has a half-life of 29 years. How long does it take for 80 percent of a sample of strontium-90 to decay?

(A) 21 years
(B) 38 years
(C) 67 years
(D) 96 years

16. After 44 minutes, a sample of $^{44}_{19}$K is found to have decayed to 25 percent of the original amount present. What is the half-life of $^{44}_{19}$K ?

(A) 11 minutes
(B) 22 minutes
(C) 44 minutes
(D) 66 minutes

Free Response Questions

1.
$$A + 2B \rightarrow 2C$$

The following results were obtained in experiments designed to study the rate of the reaction above:

Experiment	Initial Concentration (mol/L)		Initial Rate of Disappearance of A (M/sec)
	[A]	[B]	
1	0.05	0.05	3.0×10^{-3}
2	0.05	0.10	6.0×10^{-3}
3	0.10	0.10	1.2×10^{-2}
4	0.20	0.10	2.4×10^{-2}

(a) Determine the order of the reaction with respect to each of the reactants, and write the rate law for the reaction.

(b) Calculate the value of the rate constant, k, for the reaction. Include the units.

(c) If another experiment is attempted with [A] and [B], both 0.02-molar, what would be the initial rate of disappearance of A?

(d) The following reaction mechanism was proposed for the reaction above:
$$A + B \rightarrow C + D$$
$$D + B \rightarrow C$$

   (i) Show that the mechanism is consistent with the balanced reaction.

   (ii) Show which step is the rate-determining step, and explain your choice.

2.
$$2\,NO(g) + Br_2(g) \rightarrow 2\,NOBr(g)$$

The following results were obtained in experiments designed to study the rate of the reaction above:

Experiment	Initial Concentration (mol/L)		Initial Rate of Appearance of NOBr (M/sec)
	[NO]	[Br$_2$]	
1	0.02	0.02	9.6×10^{2}
2	0.04	0.02	3.8×10^{1}
3	0.02	0.04	1.9×10^{1}

(a) Write the rate law for the reaction.

(b) Calculate the value of the rate constant, k, for the reaction. Include the units.

(c) In experiment 2, what was the concentration of NO remaining when half of the original amount of Br_2 was consumed?

(d) Which of the following reaction mechanisms is consistent with the rate law established in (a)? Explain your choice.

   I. $NO + NO \rightleftharpoons N_2O_2$ (fast)

       $N_2O_2 + Br_2 \rightarrow 2\,NOBr$ (slow)

   II. $Br_2 \rightarrow Br + Br$ (slow)

       $2(NO + Br \rightarrow NOBr)$ (fast)

3.
$$N_2O_5(g) \rightarrow 4NO_2(g) + O_2(g)$$

Dinitrogen pentoxide gas decomposes according to the equation above. The first-order reaction was allowed to proceed at 40°C and the data below were collected.

$[N_2O_5]$ (M)	Time (min)
0.400	0.0
0.289	20.0
0.209	40.0
0.151	60.0
0.109	80.0

(a) Calculate the rate constant for the reaction using the values for concentration and time given in the table. Include units with your answer.
(b) After how many minutes will $[N_2O_5]$ be equal to 0.350 M?
(c) What will be the concentration of N_2O_3 after 100 minutes have elapsed?
(d) Calculate the initial rate of the reaction. Include units with your answer.
(e) What is the half-life of the reaction?

4.
$$2A + B \rightarrow C + D$$

The following results were obtained in experiments designed to study the rate of the reaction above:

Experiment	Initial Concentration (moles/L)		Initial Rate of Formation of D (M/min)
	[A]	[B]	
1	0.10	0.10	1.5×10^3
2	0.20	0.20	3.0×10^3
3	0.20	0.40	6.0×10^3

(a) Write the rate law for the reaction.
(b) Calculate the value of the rate constant, k, for the reaction. Include the units.
(c) If experiment 2 goes to completion, what will be the final concentration of D? Assume that the volume is unchanged over the course of the reaction and that no D was present at the start of the experiment.
(d) Which of the following possible reaction mechanisms is consistent with the rate law found in (a)?

   I. $A + B \rightarrow C + E$ (slow)

       $A + E \rightarrow D$ (fast)

  II. $B \rightarrow C + E$ (slow)

       $A + E \rightarrow F$ (fast)

       $A + F \rightarrow D$ (fast)

(e) Calculate the half-life of reactant B.

5.
$$A(g) + B(g) \rightarrow C(g)$$

The reaction above is second order with respect to A and zero order with respect to B. Reactants A and B are present in a closed container. Predict how each of the following changes to the reaction system will affect the rate and rate constant and explain why.

(a) More gas A is added to the container.
(b) More gas B is added to the container.
(c) The temperature is increased.
(d) An inert gas D is added to the container.
(e) The volume of the container is decreased.

6. Use your knowledge of kinetics to answer the following questions. Justify your answers.

(a)

The two lines in the diagram above show different reaction pathways for the same reaction. Which of the two lines shows the reaction when a catalyst has been added?

(b)

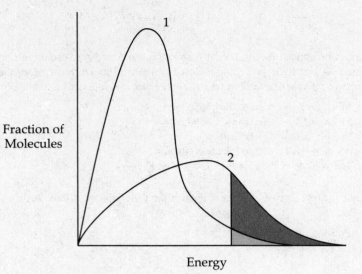

Which of the two lines in the energy distribution diagram shows the conditions at a higher temperature?

(c)

Which of the two lines in the diagram above shows the relationship of ln[A] to time for a first order reaction with the following rate law?

$$Rate = k[A]$$

(d)

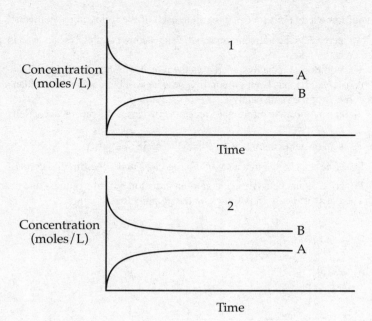

Concentration
(moles/L)

1

A
B

Time

Concentration
(moles/L)

2

B
A

Time

Which of the two graphs above shows the changes in concentration over time for the following reaction?

$$A \rightarrow B$$

(e)

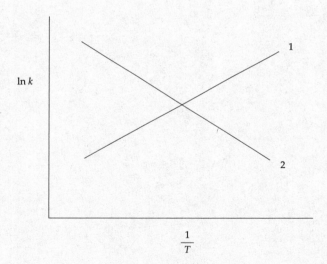

ln k

1

2

$\dfrac{1}{T}$

Which of the two lines in the diagram above shows the relationship of ln k to $\dfrac{1}{T}$ for a reaction? How is the slope of the line related to the activation energy for the reaction?

7. Use your knowledge of kinetics to explain each of the following statements:

   (a) An increase in the temperature at which a reaction takes place causes an increase in reaction rate.

   (b) The addition of a catalyst increases the rate at which a reaction will take place.

   (c) A catalyst that has been ground into powder will be more effective than a solid block of the same catalyst.

   (d) Increasing the concentration of reactants increases the rate of a reaction.

8. Neon-23 is an unstable nuclide with a half-life of 38 seconds.

   (a) Describe the change that a sample of neon-23 undergoes in 38 seconds.

   (b) Describe alpha, beta (β^-), and gamma radiation in terms of mass and charge.

   (c) The decay of neon-23 is shown on the graph below.

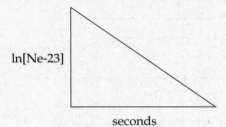

   ln[Ne-23]

   seconds

   (i)   What is the order of the nuclear reaction?
   (ii) What is represented by the slope of the graph?
   (iii) What are the units of the rate constant for this reaction?

CHAPTER 6 ANSWERS AND EXPLANATIONS

Multiple-Choice Questions

1. **A** In a first-order reaction, the exponents of all the reactants present in the rate law add up to 1.

    Rate = $k[A]^1$

2. **D** The exponent for B in this rate law is 2, so the reaction is second order with respect to B.

    Rate = $k[A]^2[B]^2$

3. **C** In a third-order reaction, the exponents of all the reactants present in the rate law add up to 3.

    Rate = $k[A]^2[B]^1$

4. **D** Let's say that [A] = [B] = 1. Then Rate = k.

    Now if we double [A] and [B], that is, we make [A] = [B] = 2, here's what we get for each of the answer choices.

(A) Rate	=	$k(2)$	= $2k$
(B) Rate	=	$k(2)^2$	= $4k$
(C) Rate	=	$k(2)(2)$	= $4k$
(D) Rate	=	$k(2)^2(2)$	= $8k$

    So in (D), the rate increases by a factor of 8.

5. **A** Let's say that [A] = [B] = 1. Then Rate = k.

    Now if we double [A] and [B], that is, we make [A] = [B] = 2, here's what we get for each of the answer choices.

(A) Rate	=	$k(2)$	= $2k$
(B) Rate	=	$k(2)^2$	= $4k$
(C) Rate	=	$k(2)(2)$	= $4k$
(D) Rate	=	$k(2)^2(2)$	= $8k$

    So in (A), the rate increases by a factor of 2.

6. **B** We need to find the rate law that is independent of changes in [A].

The only choice listed that does not include [A] is choice (B), Rate = $k[B]^2$.

7. **C** C is created and used up in the reaction, so it will not be present in the balanced equation.

8. **B** For every two NO molecules that form, two NOCl molecules must disappear, so NO is appearing at the same rate that NOCl is disappearing. Choices (D) and (E) are wrong because for every mole of Cl_2 that forms, two moles of NOCl are disappearing, so Cl_2 is appearing at *half* the rate that NOCl is disappearing.

9. **D** From the rate law given in the question (Rate = $k[H_2][I_2]$), we can see that increasing the concentration of H_2 will increase the rate of reaction. The rate constant, k, is not affected by changes in the concentration of the reactants.

10. **A** When temperature increases, the rate constant increases to reflect the fact that more reactant molecules are likely to have enough energy to react at any given time. So both the rate constant and the rate of reaction will increase.

11. **C** From a comparison of experiments 1 and 2, when [B] is doubled while [A] is held constant, the rate doubles. That means that the reaction is first order with respect to B.

From a comparison of experiments 2 and 3, when [A] is doubled while [B] is held constant, the rate doesn't change. That means that the reaction is zero order with respect to A and that A will not appear in the rate law.

So the rate law is Rate = $k[B]$.

12. **C** An increase in the concentration of A will increase the rate, as shown in the rate law for the reaction, so (I) is correct. Reactant B is not included in the rate law, so an increase in the concentration of B will not affect the rate; therefore, (II) is wrong. An increase in temperature causes more collisions with greater energy among reactants and always increases the rate of a reaction, so (III) is correct.

13. **D** From a comparison of experiments 1 and 2, when [B] is doubled while [A] is held constant, the rate doubles. That means that the reaction is first order with respect to B.

From a comparison of experiments 2 and 3, when both [A] and [B] are doubled, the rate increases by a factor of 4. We would expect the rate to double based on the change in B; because the rate is in fact multiplied by 4, the doubling of A must also change the rate by a factor of 2, so the reaction is also first order with respect to A.

So the rate law is Rate = k[A][B].

14. **A** The key to this question is to recognize that reactant A is disappearing with a characteristic half-life. This is a signal that the reaction is first order with respect to A. So the rate law must be Rate = k[A].

15. **C** Make a chart. Always start at time = 0. We're looking for the time it takes for 80 percent of the stuff to decay. That's the time when 20 percent of the stuff remains.

Half-Lives	Time	Stuff
0	0	100%
1	29 years	50%
2	58 years	25%
3	87 years	12.5%

It takes between 2 and 3 half-lives to get to the point when 20 percent remains. The correct answer is the only choice between 58 years and 87 years.

16. **B** Make a chart. Always start at time = 0.

Half-Lives	Time	Stuff
0	0	100%
1	X	50%
2	44 min.	25%

It takes two half-lives for the amount of $^{44}_{19}K$ to decrease to 25 percent. If two half-lives takes 44 minutes, one half-life must be 22 minutes.

Free Response Questions

1. (a) When we compare the results of experiments 3 and 4, we see that when [A] doubles, the rate doubles, so the reaction is first order with respect to A.

    When we compare the results of experiments 1 and 2, we see that when [B] doubles, the rate doubles, so the reaction is first order with respect to B.

    Rate = k[A][B]

    (b) Use the values from experiment 3, just because they look the simplest.

    $$k = \frac{Rate}{[A][B]} = \frac{\left(1.2 \times 10^{-2} \ M/sec\right)}{\left(0.10 \ M\right)\left(0.10 \ M\right)} = 1.2 \ M^{-1}sec^{-1} = 1.2 \ \text{L/mol-sec}$$

    (c) Use the rate law.

    Rate = k[A][B]

    Rate = $(1.2 \ M^{-1}sec^{-1})(0.02 \ M)(0.02 \ M) = 4.8 \times 10^{-4} \ M/sec$

    (d) (i)   A + B → C + D

        D + B → C

        The two reactions add up to

        A + 2 B + D → 2 C + D

        D's cancel, and we're left with the balanced equation.

        A + 2 B → 2 C

        (ii) A + B → C + D   (slow)

        D + B → C   (fast)

        The first part of the mechanism is the slow, rate-determining step because its rate law is the same as the experimentally determined rate law.

2. (a) When we compare the results of experiments 1 and 2, we see that when [NO] doubles, the rate quadruples, so the reaction is second order with respect to NO.

When we compare the results of experiments 1 and 3, we see that when $[Br_2]$ doubles, the rate doubles, so the reaction is first order with respect to Br_2.

Rate = $k[NO]^2[Br_2]$

(b) Use the values from experiment 1, just because they look the simplest.

$$k = \frac{\text{Rate}}{[NO]^2[Br_2]} = \frac{(9.6 \times 10^{-2} \ M/\text{sec})}{(0.02 \ M)^2 (0.02 \ M)} = 1.2 \times 10^4 \ M^{-2}\text{sec}^{-1} =$$

1.2×10^4 L^2/mol^2-sec

(c) In experiment 2, we started with $[Br_2]$ = 0.02 M, so 0.01 M was consumed.

From the balanced equation, 2 moles of NO are consumed for every mole of Br_2 consumed. So 0.02 M of NO are consumed.

$[NO]$ remaining = 0.04 M – 0.02 M = 0.02 M

(d) Choice (I) agrees with the rate law.

NO + NO $\leftrightarrow$ N$_2$O$_2$ (fast)

N$_2$O$_2$ + Br$_2$ $\rightarrow$ 2 NOBr   (slow)

The slow step is the rate-determining step, with the following rate law:

Rate = $k[N_2O_2][Br_2]$

We can replace the intermediate (N$_2$O$_2$) by assuming that the first step reaches equilibrium instantaneously and remembering that at equilibrium, the rates of the forward and reverse reactions are equal.

$k_f[NO]^2 = k_r[N_2O_2]$

Now solve for $[N_2O_2]$.

$[N_2O_2] = \dfrac{k_f}{k_r}[NO]^2$

Now we can substitute the rate law for the rate determining step.

Rate = $k\dfrac{k_f}{k_r}[NO]^2[Br_2]$

This matches the experimentally determined rate law. By the way, the mechanism in choice (II) would have a rate law of Rate = $k[Br_2]$.

3. (a) Use the first-order rate law and insert the first two lines from the table.

$$\ln[N_2O_5]_t - \ln[N_2O_5]_o = -kt$$

$$\ln(0.289) - \ln(0.400) = -k(20.0 \text{ min})$$

$$-.325 = -k(20.0 \text{ min})$$

$$k = 0.0163 \text{ min}^{-1}$$

(b) Use the first-order rate law.

$$\ln[N_2O_5]_t - \ln[N_2O_5]_o = -kt$$

$$\ln(0.350) - \ln(0.400) = -(0.0163 \text{ min}^{-1})t$$

$$-0.134 = -(0.0163 \text{ min}^{-1})t$$

$$t = 8.19 \text{ min}$$

(c) Use the first-order rate law.

$$\ln[N_2O_5] - \ln[N_2O_5]_o = -kt$$

$$\ln[N_2O_5] - \ln(0.400) = -(0.0163 \text{ min}^{-1})(100 \text{ min})$$

$$\ln[N_2O_5] + 0.916 = 1.63$$

$$\ln[N_2O_5] = -2.55$$

$$[N_2O_5] = e^{-2.55} \, M = 0.078 \, M$$

(d) For a first-order reaction, Rate $= k[N_2O_5] = (0.0163 \text{ min}^{-1})((0.400 \, M) = 0.0652 \, M/\text{min})$

(e) You can see from the numbers on the table that the half-life is slightly over 40 min. To calculate it exactly, use the formula

$$\text{Half-life} = \frac{0.693}{k} = \frac{0.693}{0.0163 \text{ min}^{-1}} = 42.5 \text{ min}$$

4. (a) When we compare the results of experiments 2 and 3, we see that when [B] doubles, the rate doubles, so the reaction is first order with respect to B.

Experiments 2 and 3 prove that the rate must double when [B] doubles. Knowing this, we can see that when the rate doubles from experiment 1 to 2, it must be because of B, and the change in A has no effect. So the reaction is zero order with respect to A.

Rate $= k[B]$

(b)   Take the values from experiment 1.

$$k = \frac{\text{Rate}}{[\text{B}]} = \frac{\left(1.5 \times 10^{-3} \; M/\text{min}^{-1}\right)}{\left(0.10 \; M\right)} = 1.5 \times 10^{-2} \, \text{min}^{-1}$$

(c)   In experiment 2, there are equal concentrations of A and B, but A is consumed twice as fast, so A is the limiting reagent.

0.2 M of A is consumed.

According to the balanced equation, for every 2 moles of A consumed, 1 mole of D is produced.

So if 0.2 M of A is consumed, 0.1 M of D is produced.

(d)   The reaction mechanism in choice (II) is consistent with the rate law.

B → C + E   (slow)

A + E → F   (fast)

A + F → D   (fast)

The slow, rate-determining step gives the proper rate law:

Rate = k[B]

By the way, choice (I) gives a rate law of Rate = k[A][B]

(e)   The reaction is first order with respect to B, so use the half-life formula.

$$\text{Half-Life} = \frac{0.693}{k} = \frac{0.693}{1.5 \times 10^{2} \, \text{min}^{-1}} = 46 \, \text{min}$$

5.   (a)   The rate of the reaction will increase because the rate depends on the concentration of A as given in the rate law: Rate = k[A]^2.

The rate constant is independent of the concentration of the reactants and will not change.

(b)   The rate of the reaction will not change. If the reaction is zero order with respect to B, then the rate is independent of the concentration of B.

The rate constant is independent of the concentration of the reactants and will not change.

(c)   The rate of the reaction will increase with increasing temperature because the rate constant increases with increasing temperature.

The rate constant increases with increasing temperature because at a higher temperature more gas molecules will collide with enough energy to overcome the activation energy for the reaction.

(d)   Neither the rate nor the rate constant will be affected by the addition of an inert gas.

(e)   The rate of the reaction will increase because decreasing the volume of the container will increase the concentration of A. Rate = $k[A]^2$

The rate constant is independent of the concentration of the reactants and will not change.

6.   (a)   Line 2 is the catalyzed reaction. Adding a catalyst lowers the activation energy of the reaction, making it easier for the reaction to occur.

(b)   Line 2 shows the higher temperature distribution. At a higher temperature, more of the molecules will be at higher energies, causing the distribution to flatten out and shift to the right.

(c)   Line 1 is correct. ln[reactant] for a first order reaction changes in a linear fashion over time, as shown in the following equation.

$$\ln[A]_t = -kt + \ln[A]_o$$

$$y = mx + b$$

Notice the similarity to the slope-intercept form for a linear equation.

(d)   Graph 1 is correct, showing a decrease in the concentration of A as it is consumed in the reaction, and a corresponding increase in the concentration of B as it is produced.

(e)   Line 2 is correct. The rate constant k increases with increasing temperature, so the line will slope downward when graphed against $\frac{1}{T}$. The slope of the line is proportional to the activation energy (slope $= -\frac{E_a}{R}$).

7. (a) An increase in temperature means an increase in the energy of the molecules present. If the molecules have more energy, then more of them will collide more often with enough energy to overcome the activation energy required for a reaction to take place, causing the reaction to proceed more quickly.

    (b) A catalyst offers a reaction an alternative pathway with a lower activation energy. If the activation energy is lowered, then more molecular collisions will occur with enough energy to overcome the activation energy, causing the reaction to proceed more quickly.

    (c) The effectiveness of a solid catalyst depends on the surface area of the catalyst that is exposed to the reactants. Grinding a solid into powder greatly increases its surface area.

    (d) Increasing the concentration of reactants crowds the reactants more closely together, making it more likely that they will collide with one another. The more collisions that occur, the more likely that collisions that will result in reactions will occur.

8. (a) The half-life of neon-23 is 38 seconds. That means that half of any sample of neon-23 will be converted into sodium-23 after 38 seconds.

    (b) Alpha particles have the mass (4) and the charge (+2) of a helium nucleus.

         Beta (β^-) particles have insignificant mass and the charge (–1) of an electron.

         Gamma radiation has neither mass nor charge. Gamma radiation is electromagnetic radiation.

    (c) (i)   This nuclear reaction is first order. The graph of ln[Ne-23] versus time is linear.

         (ii) The slope is the negative of the rate constant, k.

         (iii) Units are sec^{-1}.

Chapter 7

Big Idea #5: Laws of Thermodynamics and Changes in Matter

The laws of thermodynamics describe the essential role of energy and explain and predict the direction of changes in matter.

ENERGY VS TEMPERATURE

Students often make the mistake of confusing heat and temperature. Although the two terms are interchangeably used in everyday language, they mean two different things. Temperature is a measurement of the average kinetic energy of a substance. This can be best seen with the Boltzmann Distribution, which plots the number of particles versus the kinetic energy of the particle.

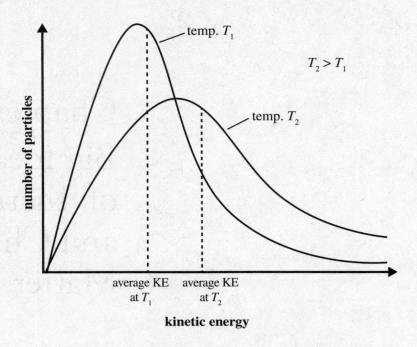

The units of temperature are °F or °R in English units, or °C or K in Metric units.

In contrast, heat is the total energy due to the molecular motions in a substance. The unit of heat is the joule in metric units. Comparing the two, it is important to remember that temperature is not energy, but a representation of the kinetic energy of the molecules.

ENERGY TRANSFER

Suppose we had two objects of different temperatures, one at 100°C and the other at 0°C, and we put the two in contact with one another. Spontaneously (i.e. it happens on its own), heat will transfer from the object with the higher temperature to the one with the lower temperature. Over time, the colder object will exhibit a temperature rise, indicating that energy has transferred from the warmer object. This is an example of heat transfer.

Now, suppose we have a container that is filled with liquid. The container is built in such a way that a paddle wheel mounted with a rotating shaft can mix the liquid inside. If someone outside of the container is turning the shaft, work, which is a form of energy, is being performed on the liquid. Thus, energy from outside the container is being transferred to the liquid inside the container (this can actually be verified because the temperature of the liquid will rise as the paddle wheel continues to turn). This is an example of energy transfer through work.

FIRST AND SECOND LAWS OF THERMODYNAMICS

The first law of thermodynamics says that the energy of the universe is constant. Energy can be neither created nor destroyed, so while energy can be *converted* in a chemical process, the total energy remains constant.

The second law of thermodynamics says that if a process is spontaneous in one direction, then it can't be spontaneous in the reverse direction, and that the entropy of the universe always increases during spontaneous reactions.

STATE FUNCTIONS

Enthalpy change (ΔH), entropy change (ΔS), and free-energy change (ΔG) are **state functions.** That means they all depend only on the change between the initial and final states of a system, not on the process by which the change occurs. For a chemical reaction, this means that the thermodynamic state functions are independent of reaction pathway; for instance, the addition of a catalyst to a reaction will have no effect on the overall energy or entropy change of the reaction.

Standard State Conditions

- All gases are at 1 atmosphere pressure.
- All liquids are pure.
- All solids are pure.
- All solutions are at 1-molar (1 M) concentration.
- The energy of formation of an element in its normal state is defined as zero.
- The temperature used for standard state values is almost invariably room temperature: 25°C (298 K). Standard state values can be calculated for other temperatures, however.

Standard State Conditions

When the values of thermodynamic quantities are given on the test, they are almost always given for standard state conditions. A thermodynamic quantity under standard state conditions is indicated by the little superscript circle, so the following is true under standard state conditions:

$$\Delta H = \Delta H°$$
$$\Delta S = \Delta S°$$
$$\Delta G = \Delta G°$$

HEAT OF FORMATION, $\Delta H°_f$

Heat of formation is the change in energy that takes place when one mole of a compound is formed from its component pure elements under standard state conditions. Heat of formation is almost always calculated at a temperature of 25°C (298 K).

Remember: $\Delta H°_f$ for a pure element is defined as zero.

- If $\Delta H°_f$ for a compound is negative, energy is released when the compound is formed from pure elements, and the product is *more* stable than its constituent elements. That is, the process is exothermic.

- If $\Delta H°_f$ for a compound is positive, energy is absorbed when the compound is formed from pure elements, and the product is *less* stable than its constituent elements. That is, the process is endothermic.

If the $\Delta H°_f$ values of the products and reactants are known, ΔH for a reaction can be calculated.

$$\Delta H° = \Sigma \Delta H°_f \text{ products} - \Sigma \Delta H°_f \text{ reactants}$$

Let's find $\Delta H°$ for the reaction below.

$$2 \, CH_3OH(g) + 3 \, O_2(g) \rightarrow 2 \, CO_2(g) + 4 \, H_2O(g)$$

Compound	$\Delta H°$ (kJ/mol)
$CH_3OH(g)$	−201
$O_2(g)$	0
$CO_2(g)$	−394
$H_2O(g)$	−242

$\Delta H° = \Sigma \Delta H°_f \text{ products} - \Sigma \Delta H°_f \text{ reactants}$
$\Delta H° = [(2)(\Delta H°_f CO_2) + (4)(\Delta H°_f H_2O)] - [(2)(\Delta H°_f CH_3OH) + (3)(\Delta H°_f O_2)]$
$\Delta H° = [(2)(−394 \text{ kJ}) + (4)(−242 \text{ kJ})] - [(2)(−201 \text{ kJ}) + (3)(0 \text{ kJ})]$
$\Delta H° = (−1,756 \text{ kJ}) - (−402 \text{ kJ})$
$\Delta H° = −1,354 \text{ kJ}$

BOND ENERGY

Bond energy is the energy required to break a bond. Because the breaking of a bond is an endothermic process, bond energy is always a positive number. When a bond is formed, energy equal to the bond energy is released.

$$\Delta H° = \Sigma \text{ Bond energies of bonds broken} - \Sigma \text{ Bond energies of bonds formed}$$

The bonds broken will be the reactant bonds, and the bonds formed will be the product bonds.

Let's find $\Delta H°$ for the reaction below.

$$2 \, H_2(g) + O_2(g) \rightarrow 2 \, H_2O(g)$$

Bond	Bond Energy (kJ/mol)
H–H	436
O–O	499
O–H	463

$\Delta H° = \Sigma$ Bond energies of bonds broken $- \Sigma$ Bond energies of bonds formed

$\Delta H° = [(2)(H–H) + (1)(O=O)] - [(4)(O–H)]$

$\Delta H° = [(2)(436 \text{ kJ}) + (1)(499 \text{ kJ})] - [(4)(463 \text{ kJ})]$

$\Delta H° = (1{,}371 \text{ kJ}) - (1{,}852 \text{ kJ})$

$\Delta H° = -481 \text{ kJ}$

THERMODYNAMICS OF PHASE CHANGE

Naming the Phase Changes

Phase changes occur because of changes in temperature and/or pressure. You should be able to read the phase diagram and see how a change in pressure or temperature will affect the phase of a substance.

Solid to *liquid*	—	**Melting**
Liquid to *solid*	—	**Freezing**
Liquid to *gas*	—	**Vaporization**
Gas to *liquid*	—	**Condensation**
Solid to *gas*	—	**Sublimation**
Gas to *solid*	—	**Deposition**

Some particles in a liquid or solid will have enough energy to break away from the surface and become gaseous. The pressure exerted by these molecules as they escape from the surface is called the **vapor pressure**. When the liquid or solid phase of a substance is in equilibrium with the gas phase, the pressure of the gas will be equal to the vapor pressure of the substance. As temperature increases, the vapor pressure of a liquid will increase. When the vapor pressure of a liquid increases to the point where it is equal to the surrounding atmospheric pressure, the liquid boils.

Heat of Fusion

The **heat of fusion** is the energy that must be put into a solid to melt it. This energy is needed to overcome the forces holding the solid together. Alternatively, the heat of fusion is the heat given off by a substance when it freezes. The intermolecular forces within a solid are more stable and, therefore, have lower energy than the forces within a liquid, so energy is released in the freezing process.

Heat of Vaporization

The **heat of vaporization** is the energy that must be put into a liquid to turn it into a gas. This energy is needed to overcome the forces holding the liquid together. Alternatively, the heat of vaporization is the heat given off by a substance when it condenses. Intermolecular forces become stronger when a gas condenses; the gas becomes a liquid, which is more stable, and energy is released.

As heat is added to a substance in equilibrium, the temperature of the substance can increase or the substance can change phases, but both changes cannot occur simultaneously.

Phase Diagram for Water

You should recognize how the phase diagram for water differs from the phase diagram for most other substances.

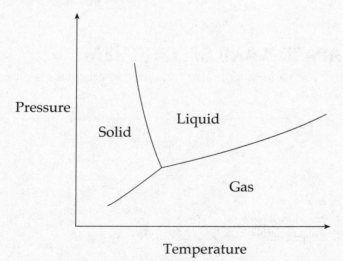

In the phase diagram for substances other than water, the solid-liquid equilibrium line slopes upward. In the phase diagram for water, the solid-liquid equilibrium line slopes *downward*. What this means is that when pressure is increased, a normal substance will change from liquid to solid, but water will change from solid to liquid.

Water has this odd property because its hydrogen bonds form a lattice structure when it freezes. This forces the molecules to remain farther apart in ice than in water, making the solid phase less dense than the liquid phase. That's why ice floats on water, and this natural phenomenon preserves most marine life.

HESS'S LAW

Hess's law states that if a reaction can be described as a series of steps, then ΔH for the overall reaction is simply the sum of the ΔH values for all the steps.

For example, let's say you wanted to calculate the enthalpy change for the following reaction:

$$C_2H_2(g) + H_2O(l) \rightarrow C_2H_5OH(l) \qquad\qquad \Delta H° = ?$$

And let's say that you know the enthalpy changes for the following two reactions:

$$C_2H_2(g) + 3\ O_2(g) \rightarrow 2\ CO_2(g) + 2\ H_2O(l) \qquad \Delta H° = -1{,}411\ kJ$$

$$2\ CO_2(g) + 3\ H_2O(l) \rightarrow C_2H_5OH(l) + 3\ O_2(g) \qquad \Delta H° = +1{,}368\ kJ$$

If you add the two reactions whose enthalpy changes you know (as though they were simultaneous equations) and cancel the substances that appear on both sides, you'll get the reaction that you're looking for. This means that you can add the enthalpy changes for the reactions that you know to get the enthalpy change that you're looking for.

$$-1{,}411\ kJ + 1{,}368\ kJ = -43\ kJ$$

HEAT CAPACITY AND SPECIFIC HEAT

Heat capacity, C_p, is a measure of how much the temperature of an object is raised when it absorbs heat.

Heat Capacity

$$C_p = \frac{\Delta H}{\Delta T}$$

C_p = heat capacity
ΔH = heat added (J or cal)
ΔT = temperature change (K or °C)

An object with a large heat capacity can absorb a lot of heat without undergoing much of a change in temperature, whereas an object with a small heat capacity shows a large increase in temperature even if only a small amount of heat is absorbed.

Specific heat is the amount of heat required to raise the temperature of one gram of a substance one degree Celsius.

> ## Specific Heat
>
> $$q = mc\Delta T$$
>
> q = heat added (J or cal)
> m = mass of the substance (g or kg)
> c = specific heat
> ΔT = temperature change (K or °C)

ENTROPY

The entropy, ΔS, of a system is a measure of the randomness or disorder of the system; the greater the disorder of a system, the greater its entropy. Because zero entropy is defined as a solid crystal at 0 K, and because 0 K has never been reached experimentally, all substances that we encounter will have some positive value for entropy. Standard entropies, $\Delta S°$, are calculated at 25°C (298 K).

The standard entropy change, $\Delta S°$, that has taken place at the completion of a reaction is the difference between the standard entropies of the products and the standard entropies of the reactants.

You should be familiar with several simple rules concerning entropies.

- Liquids have higher entropy values than solids.
- Gases have higher entropy values than liquids.
- Particles in solution have higher entropy values than solids.
- Two moles of a substance have higher entropy value than one mole.

$$\Delta S° = \Sigma \; \Delta S°_{products} - \Sigma \; \Delta S°_{reactants}$$

GIBBS FREE ENERGY

The **Gibbs free energy**, or simply free energy, ΔG, of a process is a measure of the spontaneity of the process.

For a given reaction

- if ΔG is negative, the reaction is spontaneous
- if ΔG is positive, the reaction is not spontaneous
- if $\Delta G = 0$, the reaction is at equilibrium

FREE ENERGY CHANGE, ΔG

The standard free energy change, ΔG, for a reaction can be calculated from the standard free energies of formation, ΔG°_f, of its products and reactants in the same way that ΔS° was calculated.

$$\Delta G^\circ = \Sigma \ \Delta G^\circ_{f \ products} - \Sigma \ \Delta G^\circ_{f \ reactants}$$

ΔG, ΔH, AND ΔS

In general, nature likes to move toward two different and seemingly contradictory states—low energy and high disorder, so spontaneous processes must result in decreasing enthalpy or increasing entropy or both.

There is an important equation that relates spontaneity (ΔG), enthalpy (ΔH), and entropy (ΔS) to one another.

$$\Delta G^o = \Delta H^\circ - T\Delta S^\circ$$
$$T = \text{absolute temperature (K)}$$

The chart below shows how different values of enthalpy and entropy affect spontaneity.

ΔH	ΔS	T	ΔG	
−	+	Low	−	Always spontaneous
		High	−	
+	−	Low	+	Never spontaneous
		High	+	
+	+	Low	+	Not spontaneous at low temperature
		High	−	Spontaneous at high temperature
−	−	Low	−	Spontaneous at low temperature
		High	+	Not spontaneous at high temperature

You should note that at low temperature, enthalpy is dominant, while at high temperature, entropy is dominant.

VOLTAGE AND SPONTANEITY

A redox reaction will occur spontaneously if its potential has a positive value. We also know from thermodynamics that a reaction that occurs spontaneously has a negative value for free-energy change. The relationship between reaction potential and free energy for a redox reaction is given by the equation below, which serves as a bridge between thermodynamics and electrochemistry.

$$\Delta G° = -nFE°$$

$\Delta G°$ = Standard Gibbs free energy change (kJ/mol)

n = the number of moles of electrons exchanged in the reaction (mol)

F = Faraday's constant, 96,500 coulombs/mole (that is, 1 mole of electrons has a charge of 96,500 coulombs).

$E°$ = Standard reaction potential (V)

From this equation we can see a few important things. *If $E°$ is positive, $\Delta G°$ is negative and the reaction is spontaneous, and if $E°$ is negative, $\Delta G°$ is positive and the reaction is nonspontaneous.*

Look at the reaction we saw on the previous page.

$$Zn + 2\,Ag^+ \rightarrow Zn^{2+} + 2\,Ag \qquad E° = +1.56\text{ V}$$

The reaction potential is positive, so the free energy change is negative and the reaction is spontaneous.

CHAPTER 7 QUESTIONS

Multiple-Choice Questions

Questions 1-4

(A) Free energy change (ΔG)
(B) Entropy change (ΔS)
(C) Heat of vaporization
(D) Heat of fusion

1. If this has a negative value for a process, then the process occurs spontaneously.

2. This is a measure of how the disorder of a system is changing.

3. This is the energy given off when a substance condenses.

4. This is the energy taken in by a substance when it melts.

5.
$$2 \, Al(s) + 3 \, Cl_2(g) \rightarrow 2 \, AlCl_3(s)$$

   The reaction above is not spontaneous under standard conditions, but becomes spontaneous as the temperature decreases toward absolute zero. Which of the following is true at standard conditions?

   (A) ΔS and ΔH are both negative.
   (B) ΔS and ΔH are both positive.
   (C) ΔS is negative, and ΔH is positive.
   (D) ΔS is positive, and ΔH is negative.

6.

   Which of the following is true of the reaction shown in the diagram above?

   (A) The reaction is endothermic because the reactants are at a higher energy level than the products.
   (B) The reaction is endothermic because the reactants are at a lower energy level than the products.
   (C) The reaction is exothermic because the reactants are at a higher energy level than the products.
   (D) The reaction is exothermic because the reactants are at a lower energy level than the products.

7. The addition of a catalyst will have which of the following effects on a chemical reaction?

    I. The enthalpy change will decrease.
    II. The entropy change will decrease.
    III. The activation energy will decrease.

(A) I only
(B) II only
(C) III only
(D) I and II only

8. $C(s) + 2\,H_2(g) \rightarrow CH_4(g)$ $\Delta H = x$

$C(s) + O_2(g) \rightarrow CO_2(g)$ $\Delta H = y$

$H_2(g) + \dfrac{1}{2}\,O_2(g) \rightarrow H_2O(l)$ $\Delta H = z$

Based on the information given above, what is H° for the following reaction?

$CH_4(g) + 2\,O_2(g) \rightarrow CO_2(g) + 2\,H_2O(l)$

(A) $x + y + z$
(B) $x + y - z$
(C) $z + y - 2x$
(D) $2z + y - x$

9. For which of the following processes will S be positive?

    I. $NaCl(s) \rightarrow Na^+(aq) + Cl^-(aq)$
    II. $2\,H_2(g) + O_2(g) \rightarrow 2\,H_2O(g)$
    III. $CaCO_3(s) \rightarrow CaO(s) + CO_2(g)$

(A) I only
(B) II only
(C) I and II only
(D) I and III only

10. $2\,H_2(g) + O_2(g) \rightarrow 2\,H_2O(g)$

Based on the information given in the table below, what is H° for the above reaction?

Bond	Average bond energy (kJ/mol)
H–H	500
O=O	500
O–H	500

(A)   −2,000 kJ
(B)   −500 kJ
(C)   +1,000 kJ
(D)   +2,000 kJ

11. Which of the following is true of a reaction that is spontaneous at 298 K but becomes nonspontaneous at a higher temperature?

(A) $\Delta S°$ and $\Delta H°$ are both negative.
(B) $\Delta S°$ and $\Delta H°$ are both positive.
(C) $\Delta S°$ is negative, and $\Delta H°$ is positive.
(D) $\Delta S°$ is positive, and $\Delta H°$ is negative.

12. Which of the following will be true when a pure substance in liquid phase freezes spontaneously?

(A) ΔG, ΔH, and ΔS are all positive.
(B) ΔG, ΔH, and ΔS are all negative.
(C) ΔG and ΔH are negative, but ΔS is positive.
(D) ΔG and ΔS are negative, but ΔH is positive.

13.

Which point on the graph shown above corresponds to activated complex or transition state?

(A) 1
(B) 2
(C) 3
(D) 4

14.
$$C(s) + O_2(g) \rightarrow CO_2(g) \qquad \Delta H° = -390 \text{ kJ/mol}$$

$$H_2(g) + \frac{1}{2} O_2(g) \rightarrow H_2O(l) \qquad \Delta H° = -290 \text{ kJ/mol}$$

$$2\,C(s) + H_2(g) \rightarrow C_2H_2(g) \qquad \Delta H° = +230 \text{ kJ/mol}$$

Based on the information given above, what is ΔH for the following reaction?

$$C_2H_2(g) + \frac{5}{2} O_2(g) \rightarrow 2\,CO_2(g) + H_2O(l)$$

(A)   −1,300 kJ
(B)   −1,070 kJ
(C)   −840 kJ
(D)   −780 kJ

15. In which of the following reactions is entropy increasing?

    (A) $2 SO_2(g) + O_2(g) \rightarrow 2 SO_3(g)$
    (B) $CO(g) + H_2O(g) \rightarrow H_2(g) + CO_2(g)$
    (C) $H_2(g) + Cl_2(g) \rightarrow 2 HCl(g)$
    (D) $2 NO_2(g) \rightarrow 2 NO(g) + O_2(g)$

16. When pure sodium is placed in an atmosphere of chlorine gas, the following spontaneous reaction occurs.

$$2 Na(s) + Cl_2(g) \rightarrow 2 NaCl(s)$$

Which of the following statements is true about the reaction?

    I. $\Delta S > 0$
    II. $\Delta H < 0$
    III. $\Delta G > 0$

    (A) I only
    (B) II only
    (C) I and II only
    (D) II and III only

17.
$$H_2(g) + F_2(g) \rightarrow 2 HF(g)$$

Gaseous hydrogen and fluorine combine in the reaction above to form hydrogen fluoride with an enthalpy change of –540 kJ. What is the value of the heat of formation of HF(g)?

    (A) –1,080 kJ/mol
    (B) –270 kJ/mol
    (C) 270 kJ/mol
    (D) 540 kJ/mol

18.
$$H_2O(s) \rightarrow H_2O(l)$$

Which of the following is true of the reaction shown above at room temperature?

    I. ΔG is greater than zero.
    II. ΔH is greater than zero.
    III. ΔS is greater than zero.

    (A) II only
    (B) III only
    (C) I and III only
    (D) II and III only

19.
$$2\ S(s) + 3\ O_2(g) \rightarrow 2\ SO_3(g)$$
$$\Delta H = +800\ \text{kJ/mol}$$

$$2\ SO_3(g) \rightarrow 2\ SO_2(g) + O_2(g)$$
$$\Delta H = -200\ \text{kJ/mol}$$

Based on the information given above, what is ΔH for the following reaction?

$$S(s) + O_2(g) \rightarrow SO_2(g)$$

(A)   300 kJ
(B)   500 kJ
(C)   600 kJ
(D)   1,000 kJ

Free Response Questions

1.

Substance	Absolute Entropy, S (J/mol-K)	Molecular Weight
$C_6H_{12}O_6$ (s)	212.13	180
O_2 (g)	205	32
CO_2 (g)	213.6	44
H_2O (l)	69.9	18

Energy is released when glucose is oxidized in the following reaction, which is a metabolism reaction that takes place in the body.

$$C_6H_{12}O_6(s) + 6\ O_2(g) \rightarrow 6\ CO_2(g) + 6\ H_2O(l)$$

The standard enthalpy change, $\Delta H°$, for the reaction is $-2{,}801$ kJ at 298 K.

(a)   Calculate the standard entropy change, $\Delta S°$, for the oxidation of glucose.
(b)   Calculate the standard free energy change, $\Delta G°$, for the reaction at 298 K.
(c)   What is the value of K_{eq} for the reaction?
(d)   How much energy is given off by the oxidation of 1.00 gram of glucose?

2.

Bond	Average Bond Dissociation Energy (kJ/mol)
C–H	415
O=O	495
C=O	799
O–H	463

$$CH_4(g) + 2\ O_2(g) \rightarrow CO_2(g) + 2\ H_2O(g)$$

The standard free energy change, $\Delta G°$, for the reaction above is -801 kJ at 298 K.

(a)   Use the table of bond dissociation energies to find $\Delta H°$ for the reaction above.
(b)   What is the value of K_{eq} for the reaction?
(c)   What is the value of $\Delta S°$ for the reaction at 298 K?
(d)   Give an explanation for the size of the entropy change found in (c).

3.
$$N_2(g) + 3\,H_2(g) \rightleftharpoons 2\,NH_3(g)$$

The heat of formation, ΔH°_f, of $NH_3(g)$ is -46.2 kJ/mol. The free energy of formation, ΔG°_f, of $NH_3(g)$ is -16.7 kJ/mol.

(a) What are the values of ΔH and ΔG for the reaction?
(b) What is the value of the entropy change, ΔS, for the reaction above at 298 K?
(c) As the temperature is increased, what is the effect on ΔG for the reaction? How does this affect the spontaneity of the reaction?
(d) At what temperature can N_2, H_2, and NH_3 gases be maintained together in equilibrium, each with a partial pressure of 1 atm?

4.
$$2\,H_2(g) + O_2(g) \rightarrow 2\,H_2O(l)$$

The reaction above proceeds spontaneously from standard conditions at 298 K.

(a) Predict the sign of the entropy change, ΔS°, for the reaction. Explain.
(b) How would the value of ΔS° for the reaction change if the product of the reaction was $H_2O(g)$?
(c) What is the sign of ΔG° at 298 K? Explain.
(d) What is the sign of ΔH° at 298 K? Explain.

5.
$$CaO(s) + CO_2(g) \rightarrow CaCO_3(s)$$

The reaction above is spontaneous at 298 K, and the heat of reaction, ΔH°, is -178 kJ.

(a) Predict the sign of the entropy change, ΔS°, for the reaction. Explain.
(b) What is the sign of ΔG° at 298 K? Explain.
(c) What change, if any, occurs to the value of ΔG° as the temperature is increased from 298 K?
(d) As the reaction takes place in a closed container, what changes will occur in the concentration of CO_2 and the temperature?

6.
$$H_2O(l) \rightleftharpoons H_2O(g)$$

At 298 K, the value of the equilibrium constant, K, for the reaction above is 0.036.

(a) What is the sign of ΔS° for the reaction above at 298 K?
(b) What is the sign of ΔH° for the reaction above at 298 K?
(c) What is the sign of ΔG° for the reaction above at 298 K?
(d) At approximately what temperature will ΔG for the reaction be equal to zero?

CHAPTER 7 ANSWERS AND EXPLANATIONS

Multiple-Choice Questions

1. **A** A negative value for ΔG means that the process is spontaneous. A positive value for ΔG means that the process is nonspontaneous.

2. **B** Entropy is a measure of the disorder of a system. A positive value for ΔS means that a system has become more disordered. A negative value for ΔS means that a system has become more orderly.

3. **C** The heat of vaporization is the heat given off when a substance condenses. It is also the heat that must be put into a substance to make it vaporize.

4. **D** The heat of fusion is the heat that must be put into a substance to melt it. It is also the heat given off when a substance freezes.

5. **A** Remember that $\Delta G = \Delta H - T\Delta S$.

    If the reaction is spontaneous only when the temperature is very low, then ΔG is negative only when T is very small. This can happen only when ΔH is negative (which favors spontaneity) and ΔS is negative (which favors nonspontaneity). A very small value for T will eliminate the influence of ΔS.

6. **C** In an exothermic reaction, energy is given off as the products are created because the products have less potential energy than the reactants.

7. **C** The addition of a catalyst speeds up a reaction by lowering the activation energy. A catalyst has no effect on the entropy or enthalpy change of a reaction.

8. **D** The equations given above the question give the heats of formation of all the reactants and products (remember: the heat of formation of O_2, an element in its most stable form, is zero).

    $\Delta H°$ for a reaction = ($\Delta H°$ for the products) − ($\Delta H°$ for the reactants).

    First, the products:

    > From 2 H_2O, we get $2z$
    > From CO_2, we get y
    > So $\Delta H°$ for the products = $2z + y$

Now the reactants:

> From CH_4, we get x. The heat of formation of O_2 is defined to be zero, so that's it for the reactants.
>
> $\Delta H°$ for the reaction = $(2z + y) - (x) = 2z + y - x$.

9. **D** In (I), NaCl goes from a solid to aqueous particles. Aqueous particles are more disorderly than a solid, so ΔS will be positive.

In (II), we go from 3 moles of gas to 2 moles of gas. Fewer moles and less gas means less entropy, so ΔS will be negative.

In (III), we go from a solid to a solid and a gas. More moles and the production of more gas means increasing entropy, so ΔS will be positive.

10. **B** The bond energy is the energy that must be put into a bond to break it. First let's figure out how much energy must be put into the reactants to break their bonds.

To break 2 moles of H–H bonds, it takes $(2)(500)$ kJ = 1,000 kJ

To break 1 mole of O=O bonds, it takes 500 kJ.

So to break up the reactants, it takes +1,500 kJ.

Energy is given off when a bond is formed; that's the negative of the bond energy. Now let's see how much energy is given off when 2 moles of H_2O are formed.

2 moles of H_2O molecules contain 4 moles of O–H bonds, so $(4)(-500)$ kJ = $-2,000$ kJ are given off.

So the value of $\Delta H°$ for the reaction is

$(-2,000$ kJ, the energy given off$)$ + $(1,500$ kJ, the energy put in$)$ = -500 kJ.

11. **A** Remember that $\Delta G = \Delta H - T\,\Delta S$. If the reaction is spontaneous at standard temperature but becomes nonspontaneous at higher temperatures, then ΔG is negative only at lower temperatures. This can happen only when ΔH is negative (which favors spontaneity) and ΔS is negative (which favors nonspontaneity). As the value of T increases, the influence of ΔS increases, eventually making the reaction nonspontaneous.

12. **B** The process is spontaneous, so ΔG must be negative. The intermolecular forces become stronger and the substance moves to a lower energy level when it freezes, so ΔH must be negative. The substance becomes more orderly when it freezes, so ΔS must be negative.

13. **C** Point 3 represents the activated complex, which is the point of highest energy. This point is the transition state between the reactants and the products.

14. **A** The equations given on top give the heats of formation of all the reactants and products (remember: the heat of formation of O_2, an element in its most stable form, is zero).

$\Delta H°$ for a reaction = ($\Delta H°$ for the products) − ($\Delta H°$ for the reactants).

First, the products:

From CO_2, we get (2)(−390 kJ) = −780 kJ

From H_2O, we get −290 kJ

So $\Delta H°$ for the products = (−780 kJ) + (−290 kJ) = −1,070 kJ

Now the reactants:

From C_2H_2, we get +230 kJ. The heat of formation of O_2 is defined as zero, so that's it for the reactants.

$\Delta H°$ for the reaction = (−1,070 kJ) − (+230 kJ) = −1,300 kJ.

15. **D** Choice (D) is the only reaction where the number of moles of gas is increasing, going from 2 moles of gas on the reactant side to 3 moles of gas on the product side. In all the other choices, the number of moles of gas either decreases or remains constant.

16. **B** In the reaction, a solid combines with a gas to produce fewer moles of solid, so the entropy change is negative and (I) is not true. The reaction is spontaneous, so the free energy change is negative and (III) is not true. For a reaction with decreasing entropy to be spontaneous, it must be exothermic. The enthalpy change is negative for an exothermic reaction and (II) is true.

17. **B** The reaction that forms 2 moles of HF(g) from its constituent elements has an enthalpy change of −540 kJ. The heat of formation is given by the reaction that forms 1 mole from these elements, so you can just divide −540 kJ by 2 to get −270 kJ.

18. **D**   Ice melts spontaneously at room temperature, so ΔG is less than zero; therefore, (I) is not true.

In this reaction, ice melts, so heat is absorbed to break up the intermolecular forces, and ΔH is greater than zero; therefore, (II) is true.

Liquid water is less orderly than ice, so the entropy change when ice melts is positive; therefore, ΔS is greater than zero, and (III) is true.

19. **A**   You can use Hess's law. Add the two reactions together, and cancel things that appear on both sides.

$$2\ S(s) + 3\ O_2(g) \rightarrow 2\ SO_3(g) \qquad\qquad \Delta H = +800 \text{ kJ}$$

$$2\ SO_3(g) \rightarrow 2\ SO_2(g) + O_2(g) \qquad\qquad \Delta H = -200 \text{ kJ}$$

$$2\ S(s) + 3\ O_2(g) + 2\ SO_3(g) \rightarrow 2\ SO_2(g) + 2\ SO_3(g) + O_2(g)$$

This reduces to

$$2\ S(s) + 2\ O_2(g) \rightarrow 2\ SO_2(g) \qquad\qquad \Delta H = +600 \text{ kJ}$$

Now we can cut everything in half to get the equation we want.

$$S(s) + O_2(g) \rightarrow SO_2(g) \qquad\qquad \Delta H = +300 \text{ kJ}$$

Free Response Questions

1. (a) Use the entropy values in the table.

$$\Delta S^\circ = \Sigma\ S^\circ_{products} - \Sigma\ S^\circ_{reactants}$$

$$\Delta S^\circ = [(6)(213.6) + (6)(69.9)] - [(212.13) + (6)(205)] \text{ J/K}$$

$$\Delta S^\circ = 259 \text{ J/K}$$

(b) Use the equation below. Remember that enthalpy values are given in kJ and entropy values are given in J.

$$\Delta G^\circ = \Delta H^\circ - T\ \Delta S^\circ$$

$$\Delta G^\circ = (-2{,}801 \text{ kJ}) - (298)(0.259 \text{ kJ}) = -2{,}880 \text{ kJ}$$

(c) Use the equation below. Remember that the gas constant is given in terms of J.

$$\log K = \frac{\Delta G^\circ}{-2.303RT}$$

$$\log K = \frac{(-2,880,000)}{-2.303)(8.31)(298)} = 505$$

$$K = 10^{505}$$

(d) The enthalpy change of the reaction, H°, is a measure of the energy given off by 1 mole of glucose.

$$\text{Moles} = \frac{\text{grams}}{\text{MW}}$$

$$\text{Moles of glucose} = \frac{(1.00 \text{ g})}{(180 \text{ g/mol})} = 0.00556 \text{ moles}$$

$$(0.00556 \text{ mol})(2,801 \text{ kJ/mol}) = 15.6 \text{ kJ}$$

2. (a) Use the relationship below.

$$\Delta H^\circ = \Sigma \text{ Energies of the bonds broken } - \Sigma \text{ Energies of the bonds formed}$$

$$\Delta H^\circ = [(4)(415) + (2)(495)] - [(2)(799) + (4)(463)] \text{ kJ}$$

$$\Delta H^\circ = -800 \text{ kJ}$$

(b) Use the following equation. Remember that the gas constant is given in terms of J.

$$\log K = \frac{\Delta G^\circ}{-2.303RT}$$

$$\log K = \frac{(-801,000)}{(-2.303)(8.31)(298)} = 140$$

$$K = 10^{140}$$

(c) Use $\Delta G^\circ = \Delta H^\circ - T \Delta S^\circ$

Remember that enthalpy values are given in kJ and entropy values are given in J.

$$\Delta S^\circ = \frac{\Delta H - \Delta G}{T} = \frac{(-800 \text{ kJ}) - (-801 \text{ kJ})}{(298 \text{ K})}$$

$$\Delta S^\circ = 0.003 \text{ kJ/K} = 3 \text{ J/K}$$

(d) ΔS° is very small, which means that the entropy change for the process is very small. This makes sense because the number of moles remains constant, the number of moles of gas remains constant, and the complexity of the molecules remains about the same.

3. (a)   By definition, ΔH°_f and ΔG°_f for $N_2(g)$ and $H_2(g)$ are equal to zero.

$\Delta H^\circ = \Sigma \Delta H^\circ_f \text{ products} - \Sigma \Delta H^\circ_f \text{ reactants}$

$\Delta H^\circ = [(2 \text{ mol})(- 46.2 \text{ kJ/mol})] - 0 = -92.4 \text{ kJ}$

$\Delta G^\circ = \Sigma \Delta G^\circ_f \text{ products} - \Sigma \Delta G^\circ_f \text{ reactants}$

$\Delta G^\circ = [(2 \text{ mol})(-16.7 \text{ kJ/mol})] - 0 = -33.4 \text{ kJ}$

(b)   Use $\Delta G^\circ = \Delta H^\circ - T\Delta S^\circ$

Remember that enthalpy values are given in kJ and entropy values are given in J.

$$\Delta S^\circ = \frac{\Delta H - \Delta G}{T} = \frac{(-92.4 \text{ kJ}) - (-33.4 \text{ kJ})}{(298 \text{ K})}$$
$\Delta S^\circ = -0.198 \text{ kJ/K} = -198 \text{ J/K}$

(c)   Use $\Delta G = \Delta H^\circ - T\Delta S^\circ$

From (b), ΔS° is negative, so increasing the temperature increases the value of ΔG°, making the reaction less spontaneous.

(d)   Use $\Delta G = \Delta H^\circ - T\Delta S^\circ$

At equilibrium, $\Delta G = 0$

$$T = \frac{\Delta H^\circ}{\Delta S^\circ} = \frac{(-92,400 \text{ J})}{(-198 \text{ J/K})} = 467 \text{ K}$$

4. (a) ΔS° is negative because the products are less random than the reactants. That's because gas is converted into liquid in the reaction.

(b)   The value of ΔS° would increase, becoming less negative because $H_2O(g)$ is more random than water but remaining negative because the entropy would still decrease from reactants to products.

(c) ΔG° is negative because the reaction proceeds spontaneously.

(d) ΔH° must be negative at 298 K. For a reaction to occur spontaneously from standard conditions, either ΔS° must be positive or ΔH° must be negative. This reaction is spontaneous although ΔS° is negative, so ΔH° must be negative.

5. (a) $\Delta S°$ is negative because the products are less random than the reactants. That's because two moles of reactants are converted to one mole of products and gas is converted into solid in the reaction.

(b) $\Delta G°$ is negative because the reaction proceeds spontaneously.

(c) Use $\Delta G° = \Delta H° - T\Delta S°$

$\Delta G°$ will become less negative because as temperature is increased, the entropy change of a reaction becomes more important in determining its spontaneity. The entropy change for this reaction is negative, which discourages spontaneity, so increasing temperature will make the reaction less spontaneous, thus making $\Delta G°$ less negative.

(d) The concentration of CO_2 will decrease as the reaction proceeds in the forward direction and the reactants are consumed. The temperature will increase as heat is given off by the exothermic reaction.

6. (a) $\Delta S°$ is positive because the product is more random than the reactant. That's because liquid is converted into gas in the reaction.

(b) $\Delta H°$ is positive because $H_2O(g)$ is less stable than water. Energy must be put into water to overcome intermolecular forces and create water vapor.

(c) Use $\Delta G° = -2.203RT \log K$

If K is less than 1, $\log K$ will be negative, making $\Delta G°$ positive.

(d) ΔG will be equal to zero at 373 K or 100°C. ΔG will be equal to zero at equilibrium. The point at which water and $H_2O(g)$ are in equilibrium is the boiling point.

Chapter 8
Big Idea #6: Equilibrium, Acids and Bases, Titrations, and Solubility

Any bond or intermolecular attraction that can be formed can be broken. These two processes are in dynamic competition, sensitive to initial conditions and external perturbations.

THE EQUILIBRIUM CONSTANT, K_{eq}

Most chemical processes are reversible. That is, reactants react to form products, but those products can also react to form reactants.

A reaction is at equilibrium when the rate of the forward reaction is equal to the rate of the reverse reaction.

The relationship between the concentrations of reactants and products in a reaction at equilibrium is given by the equilibrium expression, also called the **law of mass action**.

The Equilibrium Expression

For the reaction

$$aA + bB \rightleftharpoons cC + dD$$

$$K_{eq} = \frac{[C]^c [D]^d}{[A]^a [B]^b}$$

1. $[A]$, $[B]$, $[C]$, and $[D]$ are molar concentrations or partial pressures at equilibrium.
2. Products are in the numerator, and reactants are in the denominator.
3. Coefficients in the balanced equation become exponents in the equilibrium expression.
4. Solids and pure liquids are not included in the equilibrium expression—only aqueous reactants and products are included.
5. Units are not given for K_{eq}.

Let's look at a few examples:

1. $HC_2H_3O_2(aq) \rightleftharpoons H^+(aq) + C_2H_3O_2^-(aq)$

$$K_{eq} = K_a = \frac{[H^+][C_2H_3O_2^-]}{[HC_2H_3O_2]}$$

This reaction shows the dissociation of acetic acid in water. All of the reactants and products are aqueous particles, so they are all included in the equilibrium expression. None of the reactants or products have coefficients, so there are no exponents in the equilibrium expression. This is the standard form of K_a, the acid dissociation constant.

2. $2H_2S(g) + 3O_2(g) \rightleftharpoons 2H_2O(g) + 2SO_2(g)$

$$K_{eq} = K_c = \frac{[H_2O]^2 [SO_2]^2}{[H_2S]^2 [O_2]^3}$$

$$K_{eq} = K_p = \frac{P^2_{H_2O} P^2_{SO_2}}{P^2_{H_2S} P^3_{O_2}}$$

All of the reactants and products in this reaction are gases, so K_{eq} can be expressed in terms of concentration (K_c, moles/liter or molarity) or in terms of partial pressure (K_p, atmospheres). In the next section, we'll see how these two different ways of looking at the same equilibrium situation are related. All of the reactants and products are included here, and the coefficients in the reaction become exponents in the equilibrium expression.

3. $CaF_2(s) \rightleftharpoons Ca^{2+}(aq) + 2\ F^-(aq)$
    $K_{eq} = K_{sp} = [Ca^{2+}][F^-]^2$

This reaction shows the dissociation of a slightly soluble salt. There is no denominator in this equilibrium expression because the reactant is a solid. Solids are left out of the equilibrium expression because the concentration of a solid is constant. There must be some solid present for equilibrium to exist, but you do not need to include it in your calculations. This form of K_{eq} is called the solubility product, K_{sp}, which we already saw in Chapter 9.

4. $NH_3(aq) + H_2O(l) \rightleftharpoons NH_4^+(aq) + OH^-(aq)$

$$K_{eq} = K_b = \frac{[NH_4^+][OH^-]}{[NH_3]}$$

This is the acid-base reaction between ammonia and water. We can leave water out of the equilibrium expression because it is a pure liquid. By pure liquid, we mean that the concentration of water is so large (about 50 molar) that nothing that happens in the reaction is going to change it significantly, so we can consider it to be constant. This is the standard form for K_b, the base dissociation constant.

Here is a roundup of the equilibrium constants you need to be familiar with for the test.

- K_c is the constant for molar concentrations.
- K_p is the constant for partial pressures.
- K_{sp} is the solubility product, which has no denominator because the reactants are solids.
- K_a is the acid dissociation constant for weak acids.
- K_b is the base dissociation constant for weak bases.
- K_w describes the ionization of water ($K_w = 1 \times 10^{-14}$).

A large value for K_{eq} means that products are favored over reactants at equilibrium, while a small value for K_{eq} means that reactants are favored over products at equilibrium.

The equilibrium constant has a lot of aliases, but they all take the same form and tell you the same thing. The **equilibrium constant** tells you the relative amounts of products and reactants at equilibrium.

KINETICS AND EQUILIBRIUM

There is a relationship between the rate constants for the forward and reverse directions of a particular reaction and the equilibrium constant for that reaction.

$$K_{eq} = \frac{k_f}{k_r}$$

K_{eq} = the equilibrium constant
k_f = the rate constant for the forward reaction
k_r = the rate constant for the reverse reaction

K_{eq} AND GASES

As we saw in the example above, the equilibrium constant for a gas phase reaction can be written in terms of molar concentrations, K_c, or partial pressures, K_p. These two forms of K can be related by the following equation, which is derived from the ideal gas law.

$$K_p = K_c (RT)^{\Delta n}$$

K_p = partial pressure constant (using atmospheres as units)
K_c = molar concentration constant (using molarities as units)
R = the ideal gas constant, 0.0821 (L-atm)/(mol-K)
T = absolute temperature (K)
Δn = (moles of product gas – moles of reactant gas)

THE REACTION QUOTIENT, *Q*

The reaction quotient is determined in exactly the same way as the equilibrium constant, but initial conditions are used in place of equilibrium conditions. The reaction quotient can be used to predict the direction in which a reaction will proceed from a given set of initial conditions.

The Reaction Quotient

For the reaction

$$aA + bB \rightleftharpoons cC + dD$$

$$Q = \frac{[C]^c [D]^d}{[A]^a [B]^b}$$

$[A]$, $[B]$, $[C]$, and $[D]$ are initial molar concentrations or partial pressures.

- If Q is less than the calculated K for the reaction, the reaction proceeds forward, generating products.
- If Q is greater than K, the reaction proceeds backward, generating reactants.
- If $Q = K$, the reaction is already at equilibrium.

K_{eq} AND MULTISTEP PROCESSES

There is a simple relationship between the equilibrium constants for the steps of a multistep reaction and the equilibrium constant for the overall reaction.

If two reactions can be added together to create a third reaction, then the K_{eq} for the two reactions can be multiplied together to get the K_{eq} for the third reaction.

If	$A + B \rightleftharpoons C$	$K_{eq} = K_1$
and	$C \rightleftharpoons D + E$	$K_{eq} = K_2$
then	$A + B \rightleftharpoons D + E$	$K_{eq} = K_1 K_2$

LE CHÂTELIER'S LAW

Le Châtelier's law says that whenever a stress is placed on a situation at equilibrium, the equilibrium will shift to relieve that stress.

Let's use the **Haber process,** which is used in the industrial preparation of ammonia, as an example.

$$N_2(g) + 3\ H_2(g) \rightleftharpoons 2\ NH_3(g) \qquad\qquad \Delta H° = -92.6\ kJ$$

Concentration

> When the concentration of a reactant or product is increased, the reaction will proceed in the direction that will use up the added substance.

If N_2 or H_2 is added, the reaction proceeds in the forward direction. If NH_3 is added, the reaction proceeds in the reverse direction.

> When the concentration of a reactant or product is decreased, the reaction will proceed in the direction that will produce more of the substance that has been removed.

If N_2 or H_2 is removed, the reaction will proceed in the reverse direction. If NH_3 is removed, the reaction will proceed in the forward direction.

Volume

When the volume in which a reaction takes place is increased, the reaction will proceed in the direction that produces more moles of gas.

When the volume for the Haber process is increased, the reaction proceeds in the reverse direction because the reactants have more moles of gas (4) than the products (2).

> When the volume in which a reaction takes place is decreased, the reaction will proceed in the direction that produces fewer moles of gas.

When the volume for the Haber process is decreased, the reaction proceeds in the forward direction because the products have fewer moles of gas (2) than the reactants (4).

If there is no gas involved in the reaction, or if the reactants and products have the same number of moles of gas, then volume changes have no effect on the equilibrium.

Temperature

When temperature is increased, the reaction will proceed in the endothermic direction.

When the temperature for the Haber process is increased, the reaction proceeds in the reverse direction because the reverse reaction is endothermic ($\Delta H°$ is positive).

When temperature is decreased, the reaction will proceed in the exothermic direction.

When the temperature for the Haber process is decreased, the reaction proceeds in the forward direction because the forward reaction is exothermic ($\Delta H°$ is negative).

Pressure

When pressure is increased, the reaction will proceed toward the side with the fewest molecules of gas.

When pressure is decreased, the reaction will proceed toward the side with the greatest number of molecules of gas.

When an inert gas, like argon, is added to the system, although the total pressure goes up, the partial pressures of the gases involved in the reaction will remain the same. Therefore, there is no effect on the equilibrium position, because the reaction quotient, Q, is determined by the individual partial pressures of the gases involved in the reaction.

ACIDS AND BASES DEFINITIONS

Arrhenius

S. A. Arrhenius defined an acid as a substance that ionizes in water and produces hydrogen ions (H^+ ions). For instance, HCl is an acid.

$$HCl \rightarrow H^+ + Cl^-$$

He defined a base as a substance that ionizes in water and produces hydroxide ions (OH^- ions). For instance, NaOH is a base.

$$NaOH \rightarrow Na^+ + OH^-$$

Brønsted-Lowry

J. N. Brønsted and T. M. Lowry defined an acid as a substance that is capable of donating a proton, which is the same as donating an H^+ ion, and they defined a base as a substance that is capable of accepting a proton.

Look at the reversible reaction below.

$$HC_2H_3O_2 + H_2O \leftrightarrow C_2H_3O_2^- + H_3O^+$$

According to Brønsted-Lowry

$HC_2H_3O_2$ and H_3O^+ are acids.

$C_2H_3O_2^-$ and H_2O are bases.

Now look at this reversible reaction.

$$NH_3 + H_2O \leftrightarrow NH_4^+ + OH^-$$

According to Brønsted-Lowry

NH_3 and OH^- are bases.

H_2O and NH_4^+ are acids.

So in each case, the species with the H^+ ion is the acid, and the same species without the H^+ ion is the base; the two species are called a **conjugate pair**. The following are the acid-base conjugate pairs in the reactions above:

$HC_2H_3O_2$ and $C_2H_3O_2^-$

NH_4^+ and NH_3

H_3O^+ and H_2O

H_2O and OH^-

Notice that water can act either as an acid or a base.

Lewis

G. N. Lewis focused on electrons, and his definitions are the most broad of the acid-base definitions. Lewis defined a base as an electron pair donor and an acid as an electron pair acceptor; according to Lewis's rule, all of the Brønsted-Lowry bases above are also Lewis bases, and all of the Brønsted-Lowry acids are Lewis acids.

The following reaction is exclusively a Lewis acid-base reaction:

$$H-\underset{\underset{H}{|}}{\overset{\overset{H}{|}}{N}}: \ + \ B\underset{\underset{Cl}{|}}{\overset{\overset{Cl}{|}}{-}}Cl \longrightarrow H_3NBCl_3$$

NH_3 is the Lewis base, donating its electron pair, and BCl_3 is the Lewis acid, accepting the electron pair.

pH

Many of the concentration measurements in acid-base problems are given to us in terms of pH and pOH.

p (anything) = –log (anything)

$pH = -\log [H^+]$
$pOH = -\log [OH^-]$
$pK_a = -\log K_a$
$pK_b = -\log K_b$

In a solution

- when $[H^+] = [OH^-]$, the solution is neutral, and pH = 7
- when $[H^+]$ is greater than $[OH^-]$, the solution is acidic, and pH is less than 7
- when $[H^+]$ is less than $[OH^-]$, the solution is basic, and pH is greater than 7

It is important to remember that *increasing* pH means *decreasing* $[H^+]$, which means that there are fewer H^+ ions floating around and the solution is *less acidic*.

Alternatively, *decreasing* pH means *increasing* [H⁺], which means that there are more H⁺ ions floating around and the solution is *more acidic*.

WEAK ACIDS

When a weak acid is placed in water, a small fraction of its molecules will dissociate into hydrogen ions (H⁺) and conjugate base ions (B⁻). Most of the acid molecules will remain in solution as undissociated aqueous particles.

The dissociation constants, K_a and K_b, are measures of the strengths of weak acids and bases. K_a and K_b are just the equilibrium constants specific to acids and bases.

Acid Dissociation Constant

$$K_a = \frac{\left[H^+ \right]\left[A^- \right]}{\left[HA \right]}$$

[H⁺]   = molar concentration of hydrogen ions (*M*)

[A⁻]   = molar concentration of conjugate base ions (*M*)

[HA] = molar concentration of undissociated acid molecules (*M*)

Base Dissociation Constant

$$K_b = \frac{\left[HA^+ \right]\left[OH^- \right]}{\left[B \right]}$$

[HA⁺] = protonated base ions (*M*)

[OH⁻] = molar concentration of hydroxide ions (*M*)

[B]   = unprotonated base molecules (*M*)

The greater the value of K_a, the greater the extent of the dissociation of the acid and the stronger the acid. The same thing goes for K_b.

If you know the K_a for an acid and the concentration of the acid, you can find the pH. For instance, let's look at 0.20-molar solution of $HC_2H_3O_2$, with $K_a = 1.8 \times 10^{-5}$.

First we set up the K_a equation, plugging in values.

$$HC_2H_3O_2 \rightarrow H^+ + C_2H_3O_2^-$$

$$K_a = \frac{[H^+][C_2H_3O_2^-]}{[HC_2H_3O_2]}$$

Therefore, the ICE (Initial, Change, Equilibrium) table for the above problem is as follows:

	$[HC_2H_3O_2]$	$[H^+]$	$[C_2H_3O_2^-]$
Initial	0.20	0.0	0.0
Change	$-x$	$+x$	$+x$
Equilibrium	$0.20 - x$	x	x

Because every acid molecule that dissociates produces one H^+ and one $C_2H_3O_2^-$,

$$[H^+] = [C_2H_3O_2^-] = x$$

and because, strictly speaking, the molecules that dissociate should be subtracted from the initial concentration of $HC_2H_3O_2$, $[HC_2H_3O_2]$ should be ($0.20\ M - x$). In practice, however, x is almost always insignificant compared with the initial concentration of acid, so we just use the initial concentration in the calculation.

$$[HC_2H_3O_2] = 0.20\ M$$

Now we can plug our values and variable into the K_a expression.

$$1.8 \times 10^{-5} = \frac{x^2}{0.20}$$

Solve for x.

$$x = [H^+] = 1.9 \times 10^{-3}$$

Now that we know $[H^+]$, we can calculate the pH.

$$pH = -\log[H^+] = -\log(1.9 \times 10^{-3}) = 2.7$$

This is the basic approach to solving many of the weak acid/base problems that will be on the test.

STRONG ACIDS

Strong acids dissociate completely in water, so the reaction goes to completion and they never reach equilibrium with their conjugate bases. Because there is no equilibrium, there is no equilibrium constant, so there is no dissociation constant for strong acids or bases.

> **Important Strong Acids**
> $$HCl, HBr, HI, HNO_3, HClO_4, H_2SO_4$$
>
> **Important Strong Bases**
> $$LiOH, NaOH, KOH, Ba(OH)_2, Sr(OH)_2$$

Because the dissociation of a strong acid goes to completion, there is no tendency for the reverse reaction to occur, which means that the conjugate base of a strong acid must be extremely weak.

Oxoacids are acids that contain oxygen. The greater the number of oxygen atoms attached to the central atom in an oxoacid, the stronger the acid. For instance, $HClO_4$ is stronger than $HClO_3$, which is stronger than $HClO_2$. That's because increasing the number of oxygen atoms that are attached to the central atom weakens the attraction that the central atom has for the H^+ ion.

It's much easier to find the pH of a strong acid solution than it is to find the pH of a weak acid solution. That's because strong acids dissociate completely, so the final concentration of H^+ ions will be the same as the initial concentration of the strong acid.

Let's look at a 0.010-molar solution of HCl.

HCl dissociates completely, so $[H^+] = 0.010 \ M$
$pH = -\log [H^+] = -\log (0.010) = -\log (10^{-2}) = 2$

So you can always find the pH of a strong acid solution directly from its concentration.

K_w

Water comes to equilibrium with its ions according to the following reaction:

$$H_2O(l) \rightleftharpoons H^+(aq) + OH^-(aq) \qquad K_w = 1 \times 10^{-14} \text{ at } 25°C.$$
$$K_w = 1 \times 10^{-14} = [H^+][OH^-]$$
$$pH + pOH = 14$$

The common ion effect tells us that the hydrogen ion and hydroxide ion concentrations for any acid or base solution must be consistent with the equilibrium for the ionization of water. That is, no matter where the H^+ and OH^- ions came from, when you multiply $[H^+]$ and $[OH^-]$, you must get 1×10^{-14}. So for any aqueous solution, if you know the value of $[H^+]$, you can find out the value of $[OH^-]$ and vice versa.

The acid and base dissociation constants for conjugates must also be consistent with the equilibrium for the ionization of water.

$$K_w = 1 \times 10^{-14} = K_a K_b$$
$$pK_a + pK_b = 14$$

So if you know K_a for a weak acid, you can find K_b for its conjugate base and vice versa.

ACID AND BASE SALTS

A salt solution can be acidic or basic, depending on the identities of the anion and cation in the salt.

- *If a salt is composed of the conjugates of a strong base and a strong acid, its solution will be neutral.*
  For example, let's look at NaCl.
  Na^+ is the conjugate acid of NaOH (a strong base), and Cl^- is the conjugate base of HCl (a strong acid).

  $$NaCl(s) \rightarrow Na^+(aq) + Cl^-(aq) \qquad\qquad pH = 7$$

  That's because neither ion will react in water because, as you may recall, the conjugates of strong acids and bases are very weak and unreactive.

- *If a salt is composed of the conjugates of a weak base and a strong acid, its solution will be acidic.*
  For example, look at NH_4Cl.
  NH_4^+ is the conjugate acid of NH_3 (a weak base), and Cl^- is the conjugate base of HCl (a strong acid).

  $$NH_4Cl(s) \rightarrow NH_4^+(aq) + Cl^-(aq) \qquad\qquad pH < 7$$

  When NH_4Cl ionizes in water, the Cl^- ions won't react at all, but the NH_4^+ will dissociate to some extent to produce some H^+ ions, as shown in the reaction below, making the solution acidic.

  $$NH_4^+ + H_2O \rightarrow NH_3 + H_3O^+$$

- If a salt is composed of the conjugates of a strong base and a weak acid, its solution will be basic.

  For instance, look at $NaC_2H_3O_2$.

  Na^+ is the conjugate acid of $NaOH$ (a strong base), and $C_2H_3O_2^-$ is the conjugate base of $HC_2H_3O_2$ (a weak acid).

  $$NaC_2H_3O_2(s) \rightarrow Na^+(aq) + C_2H_3O_2^-(aq) \qquad\qquad pH > 7$$

  When $NaC_2H_3O_2$ ionizes in water, the Na^+ ions don't react at all, but the $C_2H_3O_2^-$ ions will react with water according to the following reaction:

  $$C_2H_3O_2^-(aq) + H_2O(l) \rightarrow HC_2H_3O_2(aq) + OH^-(aq)$$

  This is an acid-base reaction, which produces some OH^- ions and increases the pH.

- If a salt is composed of the conjugates of a weak base and a weak acid, the pH of its solution will depend on the relative strengths of the conjugate acid and base of the specific ions in the salt.

Calculations with salts aren't much different from regular acid-base calculations, except that you may need to convert between K_a and K_b and between pH and pOH.

Remember:

$$K_aK_b = 1 \times 10^{-14}, \text{ and } pH + pOH = 14$$

Let's find the pH of a 0.10-molar solution of $NaC_2H_3O_2$. K_a for $HC_2H_3O_2$ is 1.8×10^{-5}. The solution will be basic, so let's find K_b for $C_2H_3O_2^-$.

$$K_b = \frac{\left(1.0 \times 10^{-14}\right)}{K_a} = \frac{\left(1.0 \times 10^{-14}\right)}{\left(1.8 \times 10^{-5}\right)} = 5.6 \times 10^{-10}$$

Therefore, the ICE table for the above equation is as follows:

	$[HC_2H_3O_2]$	$[OH^-]$	$[C_2H_3O^{2-}]$
Initial	0.10	0.0	0.0
Change	$-x$	$+x$	$+x$
Equilibrium	$0.10 - x$	x	x

Now we can use the base dissociation expression.

$$K_b = \frac{\left[HC_2H_3O_2\right]\left[OH^-\right]}{\left[C_2H_3O_2^-\right]}$$

Every $C_2H_3O_2^-$ polyatomic ion that reacts with a water molecule produces one $HC_2H_3O_2$ and one OH^-, so

$$[HC_2H_3O_2] = [OH^-] = x$$

Again, we assume that x is insignificant compared with the initial concentration of $C_2H_3O_2^-$; 0.10 M, so

$$[C_2H_3O_2^-] = (0.10\ M - x) = 0.10\ M$$

Now we can plug our values and variable into the K_b expression.

$$5.6 \times 10^{-10} = \frac{x^2}{0.10}$$

Solve for x.

$$x = [OH^-] = 7.5 \times 10^{-6}$$

Once we know the value of $[OH^-]$, we can calculate the pOH.

$$pOH = -\log [OH^-] = -\log (7.5 \times 10^{-6}) = 5.1$$

Now we use pOH to find pH.

$$pH = 14 - pOH = 14 - 5.1 = 8.9$$

BUFFERS

A **buffer** is a solution with a very stable pH. You can add acid or base to a buffer solution without greatly affecting the pH of the solution. The pH of a buffer will also remain unchanged if the solution is diluted with water or if water is lost through evaporation.

A buffer is created by placing a large amount of a weak acid or base into a solution along with its conjugate, in the form of salt. A weak acid and its conjugate base can remain in solution together without neutralizing each other. This is called the **common ion effect**.

When both the acid and the conjugate base are together in the solution, any hydrogen ions that are added will be neutralized by the base, while any hydroxide ions that are added will be neutralized by the acid, without this having much of an effect on the solution's pH.

When dealing with buffers, it is useful to rearrange the equilibrium constant to create the **Henderson-Hasselbalch equation**.

> ### The Henderson-Hasselbalch Equation
>
> $$pH = pK_a + \log \frac{\left[A^-\right]}{\left[HA\right]}$$
>
> [HA] = molar concentration of undissociated weak acid (M)
>
> [A$^-$] = molar concentration of conjugate base (M)
>
> $$pOH = pK_b + \log \frac{\left[HB^+\right]}{\left[B\right]}$$
>
> [B] = molar concentration of weak base (M)
>
> [HB$^+$] = molar concentration of conjugate acid (M)

Let's say we have a buffer solution with concentrations of 0.20 M $HC_2H_3O_2$ and 0.50 M $C_2H_3O_2^-$. The acid dissociation constant for $HC_2H_3O_2$ is 1.8×10^{-5}. Let's find the pH of the solution.

We can just plug the values we have into the Henderson-Hasselbalch equation for acids.

$$pH = pK_a + \log \frac{\left[C_2H_3O_2^-\right]}{\left[HC_2H_3O\right]}$$

$$pH = -\log(1.8 \times 10^{-5}) + \log \frac{(0.50M)}{(0.20)}$$

$$pH = -\log(1.8 \times 10^{-5}) + \log(2.5)$$

$$pH = (4.7) + (0.40) = 5.1$$

Now let's see what happens when [$HC_2H_3O_2$] and [$C_2H_3O_2^-$] are both equal to 0.20 M.

$$pH = pK_a + \log \frac{\left[C_2H_3O_2^-\right]}{\left[HC_2H_3O\right]}$$

$$pH = -\log(1.8 \times 10^{-5}) + \log \frac{(0.20M)}{(0.20M)}$$

$$pH = -\log(1.8 \times 10^{-5}) + \log(1)$$

$$pH = (4.7) + (0) = 4.7$$

Notice that when the concentrations of acid and conjugate base in a solution are the same, pH = pK_a (and pOH = pK_b). When you choose an acid for a buffer solution, it is best to pick an acid with a pK_a that is close to the desired pH. That way you can have almost equal amounts of acid and conjugate base in the solution, which will make the buffer as flexible as possible in neutralizing both added H^+ and OH^-.

POLYPROTIC ACIDS AND AMPHOTERIC SUBSTANCES

Some acids, such as H_2SO_4 and H_3PO_4, can give up more than one hydrogen ion in solution. These are called **polyprotic** acids.

Polyprotic acids are always more willing to give up their first protons than later protons. For example, H_3PO_4 gives up an H^+ ion (proton) more easily than does $H_2PO_4^-$, so H_3PO_4 is a stronger acid. In the same way, $H_2PO_4^-$ is a stronger acid than HPO_4^{2-}.

Substances that can act as either acids or bases are called **amphoteric** substances.

For instance

- $H_2PO_4^-$ can act as an acid, giving up a proton to become HPO_4^{2-}, or it can act as a base, accepting a proton to become H_3PO_4
- HSO_4^- can act as an acid, giving up a proton to become SO_4^{2-}, or it can act as a base, accepting a proton to become H_2SO_4
- H_2O can act as an acid, giving up a proton to become OH^-, or it can act as a base, accepting a proton to become H_3O^+

ANHYDRIDES

An **acid anhydride** is a substance that combines with water to form an acid. Generally, oxides of nonmetals are acid anhydrides.

$$CO_2 + H_2O \rightarrow H_2CO_3$$

$$SO_3 + H_2O \rightarrow H^+ + HSO_4^-$$

A **basic anhydride** is a substance that combines with water to form a base. Generally, oxides of metals are basic anhydrides.

$$CaO + H_2O \rightarrow Ca(OH)_2$$

$$Na_2O + H_2O \rightarrow 2\ Na^+ + 2\ OH^-$$

TITRATION

When an acid and a base are mixed, a **neutralization reaction** occurs. Neutralization reactions can be written in the following form:

$$Acid + Base \rightarrow Water + Salt$$

Neutralization reactions are generally performed by titration, where a base of known concentration is slowly added to an acid (or vice versa). The progress of a neutralization reaction can be shown in a titration curve. The diagram below shows the titration of a strong acid by a strong base.

In the diagram above, the pH increases slowly but steadily from the beginning of the titration until just before the equivalence point. The **equivalence point** is the point in the titration when exactly enough base has been added to neutralize all the acid that was initially present. Just before the equivalence point, the pH increases sharply as the last of the acid is neutralized. The equivalence point of a titration can be recognized through the use of an indicator. An indicator is a substance that changes color over a specific pH range. When choosing an indicator, it's important to make sure the equivalence point falls within the pH range for the color change.

For this titration, the pH at the equivalence point is exactly 7 because the titration of a strong acid by a strong base produces a neutral salt solution.

The following diagram shows the titration of a weak acid by a strong base:

In this diagram, the pH increases more quickly at first, then levels out into a buffer region. At the center of the buffer region is the **half-equivalence point.** At this point, enough base has been added to convert exactly half of the acid into conjugate base; here the concentration of acid is equal to the concentration of conjugate base (pH = pK_a). The curve remains fairly flat until just before the equivalence point, when the pH increases sharply. For this titration, the pH at the equivalence point is greater than 7 because the titration of a weak acid by a strong base produces a basic salt solution.

The following diagram shows the titration curve of a polyprotic acid:

For a polyprotic acid, the titration curve will have as many bumps as there are hydrogen ions to give up. The curve above has two bumps, so it represents the titration of a diprotic acid.

SOLUBILITY

Roughly speaking, a salt can be considered "soluble" if more than 1 gram of the salt can be dissolved in 100 milliliters of water. Soluble salts are usually assumed to dissociate completely in aqueous solution. Most, but not all, solids become more soluble in a liquid as the temperature is increased.

Solubility Product (K_{sp})

Salts that are "slightly soluble" and "insoluble" still dissociate in solution to some extent. The solubility product (K_{sp}) is a measure of the extent of a salt's dissociation in solution. The K_{sp} is one of the forms of the equilibrium expression. The greater the value of the solubility product for a salt, the more soluble the salt.

Solubility Product

For the reaction

$$A_a B_b(s) \rightleftharpoons a\,A^{b+}(aq) + b\,B^{a-}(aq)$$

The solubility expression is

$$K_{sp} = [A^{b+}]^a[B^{a-}]^b$$

Here are some examples:

$$CaF_2(s) \Leftrightarrow Ca^{2+}(aq) + 2\,F^-(aq) \qquad K_{sp} = [Ca^{2+}][F^-]^2$$

$$Ag_2CrO_4(s) \Leftrightarrow 2\,Ag^+(aq) + CrO_4^{2-}(aq) \qquad K_{sp} = [Ag^+]^2[CrO_4^{2-}]$$

$$CuI(s) \Leftrightarrow Cu^+(aq) + I^-(aq) \qquad K_{sp} = [Cu^+][I^-]$$

The Common Ion Effect

Let's look at the solubility expression for AgCl.

$$K_{sp} = [Ag^+][Cl^-] = 1.6 \times 10^{-10}$$

If we throw a block of solid AgCl into a beaker of water, we can tell from the K_{sp} what the concentrations of Ag^+ and Cl^- will be at equilibrium. For every unit of

AgCl that dissociates, we get one Ag^+ and one Cl^-, so we can solve the equation above as follows:

$$[Ag^+][Cl^-] = 1.6 \times 10^{-10}$$

$$(x)(x) = 1.6 \times 10^{-10}$$

$$x^2 = 1.6 \times 10^{-10}$$

$$x = [Ag^+] = [Cl^-] = 1.3 \times 10^{-5} \ M$$

So there are very small amounts of Ag^+ and Cl^- in the solution.

Let's say we add 0.10 mole of NaCl to 1 liter of the AgCl solution. NaCl dissociates completely, so that's the same thing as adding 1 mole of Na^+ ions and 1 mole of Cl^- ions to the solution. The Na^+ ions will not affect the AgCl equilibrium, so we can ignore them; but the Cl^- ions must be taken into account. That's because of the **common ion effect.**

The common ion effect says that the newly added Cl^- ions will affect the AgCl equilibrium, although the newly added Cl^- ions did not come from AgCl.

Let's look at the solubility expression again. Now we have 0.10 mole of Cl^- ions in 1 liter of the solution, so $[Cl^-]$ = 0.10 M.

$$[Ag^+][Cl^-] = 1.6 \times 10^{-10}$$

$$[Ag^+](0.10 \ M) = 1.6 \times 10^{-10}$$

$$[Ag^+] = \frac{\left(1.6 \times 10^{-10}\right)}{(0.10)} M$$

$$[Ag^+] = 1.6 \times 10^{-9} \ M$$

Now the number of Ag^+ ions in the solution has decreased drastically because of the Cl^- ions introduced to the solution by NaCl. So when solutions of AgCl and NaCl, which share a common Cl^- ion, are mixed, the more soluble salt (NaCl) can cause the less soluble salt (AgCl) to precipitate. In general, when two salt solutions that share a common ion are mixed, the salt with the lower value for K_{sp} will precipitate first.

Solubility Rules

You should have a good working knowledge of the solubilities of common salts. This is especially useful for the part in Section II in which you are asked to predict the outcome of chemical reactions.

Cations

- **Alkali Metals:** Li^+, Na^+, K^+, Rb^+, Cs^+

  All salts of the alkali metals are **soluble.**

- **Ammonium:** NH_4^+

  All ammonium salts are **soluble.**

- **Alkaline earths and Transition Metals**

  The solubility of these elements varies depending on the identity of the anion.

Anions

The following are mostly soluble:

- **Nitrate:** NO_3^-

  All nitrate salts are **soluble.**

- **Chlorate:** ClO_3^-

  All chlorate salts are **soluble.**

- **Perchlorate:** ClO_4^-

  All perchlorate salts are **soluble.**

- **Acetate:** $C_2H_3O_2^-$

  All acetate salts are **soluble.**

- **Chloride, Bromide, Iodide:** Cl^-, Br^-, I^-

  Salts containing Cl^-, Br^-, and I^- are **soluble**
  EXCEPT for those containing Ag^+, Pb^{2+}, and Hg_2^{2+}.

- **Sulfate:** SO_4^{2-}

  Sulfate salts are **soluble**
  EXCEPT for those containing Ag^+, Pb^{2+}, Hg_2^{2+}, Ca^{2+}, Sr^{2+}, and Ba^{2+}.

The following are mostly insoluble:

- **Hydroxide:** OH^-

  Hydroxide salts are **insoluble**
  EXCEPT for those containing alkali metals, which are soluble
  AND those containing Ca^{2+}, Sr^{2+}, and Ba^{2+}, which fall in the gray area of moderate solubility.

- **Carbonate:** CO_3^{2-}

  Carbonate salts are **insoluble**
  EXCEPT for those containing alkali metals and ammonium, which are soluble.

- **Phosphate:** PO_4^{3-}

  Phosphate salts are **insoluble**
  EXCEPT for those containing alkali metals and ammonium, which are soluble.

- **Sulfite:** SO_3^{2-}

  Sulfite salts are **insoluble**
  EXCEPT for those containing alkali metals and ammonium, which are soluble.

- **Chromate:** CrO_4^{2-}
  Chromate salts are **insoluble**
  EXCEPT for those containing alkali metals and ammonium, which are soluble.
- **Sulfide:** S^{2-}
  Sulfide salts are **insoluble**
  EXCEPT for those containing alkali metals, the alkaline earths, and ammonium, which are soluble.

Solubility of Gases

The lower the temperature and the higher the pressure of the gas, the more soluble the gas will be. Think of what happens when you open a bottle of warm seltzer. The gas suddenly escapes from the warm liquid when you release the pressure.

ΔG AND $\Delta G°$

The standard free energy change, $\Delta G°$, gives the spontaneity of a reaction when all the concentrations of reactants and products are in their standard state concentrations (at 1 molar). The free energy change, and thus the spontaneity, of a reaction will be different from the standard free energy change if the initial concentrations of reactants and products are not 1 molar.

The standard free energy change, $\Delta G°$, can be related to ΔG for other conditions by the following equations:

$$\Delta G = \Delta G° + RT \ln Q$$
$$\text{or}$$
$$\Delta G = \Delta G° + 2.303 RT \log Q$$

$\Delta G°$ = standard free energy change (J)
ΔG = free energy change under given initial conditions (J)
R = the gas constant, 8.31 J/mol-K
T = absolute temperature (K)
Q = the reaction quotient for the given initial conditions

Standard Free Energy Change and the Equilibrium Constant

Let's look at the equation relating ΔG and $\Delta G°$.

$$\Delta G = \Delta G° + RT \ln Q$$

At equilibrium, $\Delta G = 0$ and $Q = K$.

Knowing this, we can derive an equation relating the standard free energy change, $\Delta G°$, and the equilibrium constant, K.

$$\Delta G° = -RT \ln K$$

or

$$\Delta G° = -2.303RT \log K$$

$\Delta G°$ = the gas constant, 8.31 J/mol-K
T = absolute temperature (K)
K = the equilibrium constant

Notice that if $\Delta G°$ is negative, K must be greater than 1, and products will be favored at equilibrium. Alternatively, if $\Delta G°$ is positive, K must be less than 1, and reactants will be favored at equilibrium.

CHAPTER 8 QUESTIONS

Multiple Choice

Questions 1–3

    (A) K_c
    (B) K_p
    (C) K_a
    (D) K_w

1. This equilibrium constant uses partial pressures of gases as units.

2. This equilibrium constant always has a value of 1×10^{-14} at 25°C.

3. This equilibrium constant is used for the dissociation of an acid.

Questions 4–7

    (A)   1
    (B)   3
    (C)   7
    (D) 13

4. The pH of a solution with a pOH of 11

5. The pH of a 0.1-molar solution of HCl

6. The pH of a 0.001-molar solution of HNO_3

7. The pH of a 0.1-molar solution of NaOH

Questions 8–11

    (A) HNO_3
    (B) HCN
    (C) H_2CO_3
    (D) H_2O

8. This is a strong electrolyte.

9. This substance can act as a Lewis base.

10. A 0.1-molar solution of this substance will have the lowest pH of the substances listed.

11. This substance is formed when carbon dioxide is dissolved in water.

12. When sodium chloride is added to a saturated aqueous solution of silver chloride, which of the following precipitates would be expected to appear?

(A) Sodium
(B) Silver
(C) Sodium chloride
(D) Silver chloride

13. The solubility product, K_{sp}, of AgCl is 1.8×10^{-10}. Which of the following expressions is equal to the solubility of AgCl?

(A) $\left(1.8 \times 10^{-10}\right)^2$ molar

(B) $\dfrac{1.8 \times 10^{-10}}{2}$ molar

(C) 1.8×10^{-10} molar

(D) $\sqrt{1.8 \times 10^{-10}}$ molar

14. A 0.1-molar solution of which of the following acids will be the best conductor of electricity?

(A) H_2CO_3
(B) H_2S
(C) HF
(D) HNO_3

15. Which of the following expressions is equal to the K_{sp} of Ag_2CO_3?

(A) $K_{sp} = [Ag^+][CO_3^{2-}]$
(B) $K_{sp} = [Ag^+][CO_3^{2-}]^2$
(C) $K_{sp} = [Ag^+]^2[CO_3^{2-}]$
(D) $K_{sp} = [Ag^+]^2[CO_3^{2-}]^2$

16. If the solubility of BaF_2 is equal to x, which of the following expressions is equal to the solubility product, K_{sp}, for BaF_2?

(A) x^2
(B) $2x^2$
(C) $2x^3$
(D) $4x^3$

17. A beaker contains 50.0 ml of a 0.20 M Na_2SO_4 solution. If 50.0 ml of a 0.10 M solution of $Ba(NO_3)_2$ is added to the beaker, what will be the final concentration of sulfate ions in the solution?

(A) 0.20 M
(B) 0.10 M
(C) 0.050 M
(D) 0.025 M

18. The solubility of strontium fluoride in water is 1×10^{-3} M at room temperature. What is the value of the solubility product for SrF_2?

    (A) 2×10^{-3}
    (B) 4×10^{-6}
    (C) 2×10^{-6}
    (D) 4×10^{-9}

19. The bottler of a carbonated beverage dissolves carbon dioxide in water by placing carbon dioxide in contact with water at a pressure of 1 atm at room temperature. The best way to increase the amount of dissolved CO_2 would be to

    (A) increase the temperature and increase the pressure of CO_2.
    (B) decrease the temperature and decrease the pressure of CO_2.
    (C) decrease the temperature and increase the pressure of CO_2.
    (D) increase the temperature without changing the pressure of CO_2.

20. When 300. ml of a 0.60 M NaCl solution is combined with 200. ml of a 0.40 M $MgCl_2$ solution, what will be the molar concentration of Cl^- ions in the solution?

    (A) 0.20 M
    (B) 0.34 M
    (C) 0.68 M
    (D) 0.80 M

21. Silver hydroxide will be LEAST soluble in a solution with a pH of

    (A) 5
    (B) 7
    (C) 9
    (D) 11

22. A student added 0.10 mol of NaBr and 0.20 mol of $BaBr_2$ to 2 liters of water to create an aqueous solution. What is the minimum number of moles of $Ag(C_2H_3O_2)$ that the student must add to the solution to precipitate out all of the Br^- ions as AgBr?

    (A) 0.20
    (B) 0.30
    (C) 0.40
    (D) 0.50

23. A student added 1 liter of a 1.0 M KCl solution to 1 liter of a 1.0 M $Pb(NO_3)_2$ solution. A lead chloride precipitate formed, and nearly all of the lead ions disappeared from the solution. Which of the following lists the ions remaining in the solution in order of decreasing concentration?

    (A) $[NO_3^-] > [K^+] > [Pb^{2+}]$
    (B) $[NO_3^-] > [Pb^{2+}] > [K^+]$
    (C) $[K^+] > [Pb^{2+}] > [NO_3^-]$
    (D) $[K^+] > [NO_3^-] > [Pb^{2+}]$

24. The solubility of PbS in water is 3×10^{-14} molar. What is the solubility product constant, K_{sp}, for PbS?

(A) 9×10^{-7}
(B) 3×10^{-14}
(C) 3×10^{-28}
(D) 9×10^{-28}

25. For a particular salt, the solution process is endothermic. As the temperature at which the salt is dissolved increases, which of the following will occur?

(A) K_{sp} will increase, and the salt will become more soluble.
(B) K_{sp} will decrease, and the salt will become more soluble.
(C) K_{sp} will increase, and the salt will become less soluble.
(D) K_{sp} will not change, and the salt will become more soluble.

26. $2 HI(g) + Cl_2(g) \rightleftharpoons 2 HCl(g) + I_2(g) + energy$

A gaseous reaction occurs and comes to equilibrium as shown above. Which of the following changes to the system will serve to increase the number of moles of I_2 present at equilibrium?

(A) Increasing the volume at constant temperature
(B) Decreasing the volume at constant temperature
(C) Increasing the temperature at constant volume
(D) Decreasing the temperature at constant volume

27. A sealed isothermal container initially contained 2 moles of CO gas and 3 moles of H_2 gas. The following reversible reaction occurred:

$$CO(g) + 2 H_2(g) \rightleftharpoons CH_3OH(g)$$

At equilibrium, there was 1 mole of CH_3OH in the container. What was the total number of moles of gas present in the container at equilibrium?

(A) 1
(B) 2
(C) 3
(D) 4

28. A sample of solid potassium nitrate is placed in water. The solid potassium nitrate comes to equilibrium with its dissolved ions by the endothermic process shown below.

$$KNO_3(s) + energy \rightleftharpoons K^+(aq) + NO_3^-(aq)$$

Which of the following changes to the system would increase the concentration of K^+ ions at equilibrium?

(A) The volume of the solution is increased.
(B) The volume of the solution is decreased.
(C) Additional solid KNO_3 is added to the solution.
(D) The temperature of the solution is increased.

29. For which of the following gaseous equilibria do K_p and K_c differ the most?

(A) $2H_2(g) + O_2(g) \rightleftharpoons 2H_2O(g)$

(B) $NH_3BH_3(g) \rightleftharpoons NH_2BH_2(g) + H_2(g)$

(C) $NO(g) + O_3(g) \rightleftharpoons NO_2(g) + O_2(g)$

(D) $B_3N_3H_6(g) + 3H_2(g) \rightleftharpoons B_3N_3H_{12}(g)$

30. A 1 M solution of $SbCl_5$ in organic solvent shows no noticeable reactivity at room temperature. The sample is heated to 200°C and the following equilibrium reaction is found to occur:

$$SbCl_5 \rightleftharpoons SbCl_3 + Cl_2$$

The value of K_{eq} at 200°C was measured to be 10^{-6}. The reaction was then heated to 350°C, and the equilibrium concentration of Cl_2 was found to be 0.1 M, which of the following best approximated the value of K_{eq} at 350°C?

(A) 1.5

(B) 1.0

(C) 0.1

(D) 0.01

31. $2\, NOBr(g) \rightleftharpoons 2\, NO(g) + Br_2(g)$

The reaction above came to equilibrium at a temperature of 100°C. At equilibrium the partial pressure due to NOBr was 4 atmospheres, the partial pressure due to NO was 4 atmospheres, and the partial pressure due to Br_2 was 2 atmospheres. What is the equilibrium constant, K_p, for this reaction at 100°C?

(A) $\dfrac{1}{4}$

(B) $\dfrac{1}{2}$

(C) 1

(D) 2

32. $Br_2(g) + I_2(g) \leftrightarrow 2\, IBr(g)$

At 150°C, the equilibrium constant, K_c, for the reaction shown above has a value of 300. This reaction was allowed to reach equilibrium in a sealed container and the partial pressure due to $IBr(g)$ was found to be 3 atm. Which of the following could be the partial pressures due to $Br_2(g)$ and $I_2(g)$ in the container?

	$Br_2(g)$	$I_2(g)$
(A)	0.1 atm	0.3 atm
(B)	0.3 atm	1 atm
(C)	1 atm	1 atm
(D)	1 atm	3 atm

33. $H_2(g) + CO_2(g) \leftrightarrow H_2O(g) + CO(g)$

Initially, a sealed vessel contained only $H_2(g)$ with a partial pressure of 6 atm and $CO_2(g)$ with a partial pressure of 4 atm. The reaction above was allowed to come to equilibrium at a temperature of 700 K. At equilibrium, the partial pressure due to $CO(g)$ was found to be 2 atm. What is the value of the equilibrium constant K_p, for the reaction?

(A) $\dfrac{1}{6}$

(B) $\dfrac{1}{4}$

(C) $\dfrac{1}{3}$

(D) $\dfrac{1}{2}$

34. What is the volume of 0.05-molar HCl that is required to neutralize 50 ml of a 0.10-molar $Mg(OH)_2$ solution?

(A) 100 ml
(B) 200 ml
(C) 300 ml
(D) 400 ml

35. Which of the following best describes the pH of a 0.01-molar solution of HBrO? (For HBrO, $K_a = 2 \times 10^{-9}$)

(A) Less than or equal to 2
(B) Between 2 and 7
(C) 7
(D) Between 7 and 11

36. A 0.5-molar solution of which of the following salts will have the lowest pH?

(A) KCl
(B) $NaC_2H_3O_2$
(C) NaI
(D) NH_4Cl

37. Which of the following salts will produce a solution with a pH of greater than 7 when placed in distilled water?

(A) NaCN
(B) KCl
(C) $NaNO_3$
(D) NH_4NO_3

38. A laboratory technician wishes to create a buffered solution with a pH of 5. Which of the following acids would be the best choice for the buffer?

(A) $H_2C_2O_4$, $K_a = 5.9 \times 10^{-2}$

(B) H_3AsO_4, $K_a = 5.6 \times 10^{-3}$

(C) $H_2C_2H_3O_2$, $K_a = 1.8 \times 10^{-5}$

(D) $HOCl$, $K_a = 3.0 \times 10^{-8}$

39. Which of the following species is amphoteric?

(A) H^+

(B) CO_3^{2-}

(C) HCO_3^-

(D) H_2CO_3

40. How many liters of distilled water must be added to 1 liter of an aqueous solution of HCl with a pH of 1 to create a solution with a pH of 2?

(A) 0.1 L
(B) 0.9 L
(C) 2 L
(D) 9 L

41. A 1-molar solution of a very weak monoprotic acid has a pH of 5. What is the value of K_a for the acid?

(A) $K_a = 1 \times 10^{-10}$
(B) $K_a = 1 \times 10^{-7}$
(C) $K_a = 1 \times 10^{-5}$
(D) $K_a = 1 \times 10^{-2}$

42. The value of K_a for HSO_4^- is 1×10^{-2}. What is the value of K_b for SO_4^{2-}?

(A) $K_b = 1 \times 10^{-12}$
(B) $K_b = 1 \times 10^{-8}$
(C) $K_b = 1 \times 10^{-2}$
(D) $K_b = 1 \times 10^2$

43. How much 0.1-molar NaOH solution must be added to 100 milliliters of a 0.2-molar H_2SO_3 solution to neutralize all of the hydrogen ions in H_2SO_3?

(A) 100 ml
(B) 200 ml
(C) 300 ml
(D) 400 ml

44. Which of the following species is amphoteric?

(A) HNO_3
(B) $HC_2H_3O_2$
(C) HSO_4^-
(D) H_3PO_4

45. If 0.630 grams of HNO_3 (molecular weight 63.0) are placed in 1 liter of distilled water at 25°C, what will be the pH of the solution? (Assume that the volume of the solution is unchanged by the addition of the HNO_3.)

(A) 0.01
(B) 0.1
(C) 1
(D) 2

46. Which of the following procedures will produce a buffered solution?

I. Equal volumes of 0.5 M NaOH and 1 M HCl solutions are mixed.
II. Equal volumes of 0.5 M NaOH and 1 M $HC_2H_3O_2$ solutions are mixed.
III. Equal volumes of 1 M $NaC_2H_3O_2$ and 1 M $HC_2H_3O_2$ solutions are mixed.

(A) I only
(B) III only
(C) I and II only
(D) II and III only

47. Which of the following compounds will most likely act as a *Lewis* acid?

(A) $AlBr_3$
(B) CH_4
(C) NH_3
(D) PCl_3

48. The following titration curve shows the titration of a weak base with a strong acid:

Mol acid titrated

Which of the following values most accurately approximates the pK_b of the weak base?

(A) 9.8
(B) 8.4
(C) 7
(D) 4.2

Free Response Questions

1. The value of the solubility product, K_{sp}, for calcium hydroxide, $Ca(OH)_2$, is 5.5×10^{-6}, at 25°C.

   (a) Write the K_{sp} expression for calcium hydroxide.

   (b) What is the mass of $Ca(OH)_2$ in 500 ml of a saturated solution at 25°C?

   (c) What is the pH of the solution in (b)?

   (d) If 1.0 mole of OH^- is added to the solution in (b), what will be the resulting Ca^{2+} concentration? Assume that the volume of the solution does not change.

2. For sodium chloride, the solution process with water is endothermic.

   (a) Describe the change in entropy when sodium chloride dissociates into aqueous particles.

   (b) Two saturated aqueous NaCl solutions, one at 20°C and one at 50°C, are compared. Which one will have higher concentration? Justify your answer.

   (c) The solubility product of $Ce_2(SO_4)_3$ decreases as temperature increases. Is the solution process for this salt endothermic or exothermic? Justify your answer.

3.
$$H_2CO_3 \rightleftharpoons H^+ + HCO_3^- \qquad K_1 = 4.3 \times 10^{-7}$$
$$HCO_3^- \rightleftharpoons H^+ + CO_3^{2-} \qquad K_2 = 5.6 \times 10^{-11}$$

   The acid dissociation constants for the reactions above are given at 25°C.

   (a) What is the pH of a 0.050-molar solution of H_2CO_3 at 25°C?

   (b) What is the concentration of CO_3^{2-} ions in the solution in (a)?

   (c) How would the addition of each of the following substances affect the pH of the solution in (a)?

   (i) HCl

   (ii) $NaHCO_3$

   (iii) NaOH

   (iv) NaCl

   (d) What is the value of K_{eq} for the following reaction?

$$H_2CO_3 \rightleftharpoons 2\,H^+ + CO_3^{2-}$$

4. $N_2(g) + 3 H_2(g) \rightleftharpoons 2 NH_3(g)$ $\Delta H = -92.4$ kJ

    When the reaction above took place at a temperature of 570 K, the following equilibrium concentrations were measured:

    $[NH_3] = 0.20$ mol/L

    $[N_2] = 0.50$ mol/L

    $[H_2] = 0.20$ mol/L

    (a) Write the expression for K_c and calculate its value.

    (b) What is the value of K_p for the reaction?

    (c) Describe how the concentration of H_2 will be affected by each of the following changes to the system at equilibrium:

        (i)   The temperature is increased.

        (ii) The volume of the reaction chamber is increased.

        (iii) N_2 gas is added to the reaction chamber.

        (iv) Helium gas is added to the reaction chamber.

5. A vessel contains 500 milliliters of a 0.100-molar H_2S solution. For H_2S, $K_1 = 1.0 \times 10^{-7}$ and $K_2 = 1.3 \times 10^{-13}$.

    (a) What is the pH of the solution?

    (b) How many milliliters of 0.100-molar NaOH solution must be added to the solution to create a solution with a pH of 7?

    (c) What will be the pH when 800 milliliters of 0.100-molar NaOH has been added?

    (d) What is the value of K_{eq} for the following reaction?

        $$H_2S(aq) \leftrightarrow 2H^+(aq) + S^{2-}(aq)$$

6. A 100 milliliter sample of 0.100-molar NH_4Cl solution was added to 80 milliliters of a 0.200-molar solution of NH_3. The value of K_b for ammonia is 1.79×10^{-5}.

    (a) What is the value of pK_b for ammonia?

    (b) What is the pH of the solution described in the question?

    (c) If 0.200 grams of NaOH were added to the solution, what would be the new pH of the solution? (Assume that the volume of the solution does not change.)

    (d) If equal molar quantities of NH_3 and NH_4^+ were mixed in solution, what would be the pH of the solution?

CHAPTER 8 ANSWERS AND EXPLANATIONS

Multiple-Choice Questions

1. **B** K_p is used for gaseous reactions, and the units used are partial pressures.

2. **D** K_w is the dissociation constant for water.

    At 25°C, $K_w = [H^+][OH^-] = 1 \times 10^{-14}$.

3. **C** K_a is known as the acid dissociation constant.

4. **B**   In any aqueous solution at 25°C, pH + pOH = 14.

    So pH + 3 = 14 and pH = 3.

5. **A**   HCl is a strong acid and dissociates completely, so the hydrogen ion concentration is equal to the molarity.

    So $[H^+] = 0.1\ M$

    pH = –log[H$^+$] = –log(0.1) = 1

6. **B** HNO_3 is a strong acid and dissociates completely, so the hydrogen ion concentration is equal to the molarity.

    So $[H^+] = 0.001\ M$

    pH = –log[H$^+$] = –log(0.001) = 3

7. **D**   NaOH is a strong acid and dissociates completely, so the hydroxide ion concentration [OH$^-$] is equal to the molarity.

    So $[OH^-] = 0.1\ M$

    pOH = –log[OH$^-$] = –log(0.1) = 1

    Now, pH + pOH = 14

    So pH + 1 = 14 and pH = 13

8. **A**   A strong electrolyte is a substance that dissociates to form a lot of ions. HNO_3 is the only substance listed that ionizes completely in solution.

9.   **D**   Water can act as a Lewis base by donating an electron pair to a hydrogen ion to form H_3O^+.

10.   **A** HNO_3 is the only strong acid listed, meaning that it is the only acid listed that ionizes completely, so it will have the lowest pH. By the way, in a 0.1 M solution of a strong acid, the hydrogen ion concentration will be 0.1 M, so the pH will be equal to 1.

11.   **C** H_2CO_3, carbonic acid, is formed in the reaction between CO_2 and H_2O.

12.   **D**   Sodium chloride is much more soluble than silver chloride. Because of the common ion effect, the chloride ions introduced into the solution by sodium chloride will disrupt the silver chloride equilibrium, causing silver chloride to precipitate from the solution.

13.   **D**   The solubility of a substance is equal to its maximum concentration in solution.

For every AgCl in solution, we get one Ag^+ and one Cl^-, so the solubility of AgCl—let's call it x—will be the same as $[Ag^+]$, which is the same as $[Cl^-]$.

So for AgCl, $K_{sp} = [Ag^+][Cl^-] = 1.8 \times 10^{-10} = x^2$.

$$x = \sqrt{1.8 \times 10^{-10}}$$

14.   **D**   The best conductor of electricity (also called the strongest electrolyte) will be the solution that contains the most charged particles. HNO_3 is the only strong acid listed in the answer choices, so it is the only choice where the acid has dissociated completely in solution into H^+ and NO_3^- ions. So a 0.1-molar HNO_3 solution will contain the most charged particles and, therefore, be the best conductor of electricity.

15.   **C** K_{sp} is just the equilibrium constant without a denominator.

When Ag_2CO_3 dissociates, we get the following reaction:

$$Ag_2CO_3(s) \rightleftharpoons 2\,Ag^+ + CO_3^{2-}$$

In the equilibrium expression, coefficients become exponents, so we get

$$K_{sp} = [Ag^+]^2[CO_3^{2-}]$$

16. **D** For BaF_2, $K_{sp} = [Ba^{2+}][F^-]^2$.

For every BaF_2 that dissolves, we get one Ba^{2+} and two F^-.

So if the solubility of BaF_2 is x, then $[Ba^{2+}] = x$, and $[F^-] = 2x$

So $K_{sp} = (x)(2x)^2 = (x)(4x^2) = 4x^3$

17. **C** The Ba^{2+} ions and the SO_4^- ions will combine and precipitate out of the solution, so let's find out how many of each we have.

Moles = (molarity)(volume)
Moles of SO_4^- = (0.20 M)(0.050 L) = 0.010 mole
Moles of Ba^{2+} = (0.10 M)(0.050 L) = 0.0050 mole

To find the number of moles of SO_4^- left in the solution, subtract the moles of Ba^{2+} from the moles of SO_4^-.

0.010 mole – 0.0050 mole = 0.0050 mole

Now use the formula for molarity to find the concentration of SO_4^- ions. Don't forget to add the volumes of the two solutions.

$$\text{Molarity} = \frac{\text{moles}}{\text{liters}} = \frac{(0.0050 \text{ mol})}{(0.10 L)} = 0.050 \text{ M}$$

18. **D** Use the formula for K_{sp}. For every SrF_2 in solution, there will be one strontium ion and two fluoride ions, so $[Sr^{2+}]$ will be 1×10^{-3} M and $[F^-]$ will be 2×10^{-3} M.

$K_{sp} = [Sr^{2+}][F^-]^2$
$K_{sp} = (1 \times 10^{-3} M)(2 \times 10^{-3} M)^2$
$K_{sp} = 4 \times 10^{-9}$

19. **C** The lower the temperature, the more soluble a gas will be in water. The greater the pressure of the gas, the more soluble it will be.

20.   C   The Cl⁻ ions from the two salts will both be present in the solution, so we need to find the number of moles of Cl⁻ contributed by each salt.

Moles = (molarity)(volume)
Each NaCl produces 1 Cl⁻
Moles of Cl⁻ from NaCl = (0.60 M)(0.300 L) = 0.18 mole

Each $MgCl_2$ produces 2 Cl⁻
Moles of Cl⁻ from $MgCl_2$ = (2)(0.40 M)(0.200 L) = 0.16 mole

To find the number of moles of Cl⁻ in the solution, add the two together.

0.18 mole + 0.16 mole = 0.34 mole

Now use the formula for molarity to find the concentration of Cl⁻ ions. Don't forget to add the volumes of the two solutions.

$$\text{Molarity} = \frac{\text{moles}}{\text{liters}} = \frac{(0.34\,\text{mol})}{(0.500\,\text{L})} = 0.68 \text{ M}$$

21.   D   Silver hydroxide will be least soluble in the solution with the highest hydroxide concentration. That would be the solution with the highest pH.

22.   D   The solution contains 0.50 moles of Br⁻ ions, 0.10 from NaBr and 0.40 from $BaBr_2$ (each $BaBr_2$ provides 2 Br⁻ ions). Each Ag⁺ ion will remove 1 Br⁻ ion, so the student needs to add 0.50 moles of $Ag(C_2H_3O_2)$.

23.   A   At the start, the concentrations of the ions are as follows:

[K⁺] = 1 M
[Cl⁻] = 1 M
[Pb²⁺] = 1 M
[NO₃⁻] = 2 M

After $PbCl_2$ forms, the concentrations are as follows:

[K⁺] = 1 M
[Cl⁻] = 0.5 M
[Pb²⁺] = 0 M
[NO₃⁻] = 2 M

So from greatest to least

[NO₃⁻] > [K⁺] > [Pb²⁺]

24. **D** The solubility of a substance is equal to its maximum concentration in solution. For every PbS in solution, we get one Pb^{2+} and one S^{2-}, so the concentration of PbS, 3×10^{-14} M, will be the same as the concentrations of Pb^{2+} and S^{2-}.

$K_{sp} = [Pb^{2+}][S^{2-}]$
$K_{sp} = (3 \times 10^{-14} M)(3 \times 10^{-14} M) = 9 \times 10^{-28}$

25. **A** From Le Châtelier's law, the equilibrium will shift to counteract any stress that is placed on it. Increasing temperature favors the endothermic direction of a reaction because the endothermic reaction absorbs the added heat. So the salt becomes more soluble, increasing the number of dissociated particles, thus increasing the value of K_{sp}.

26. **D** According to Le Châtelier's law, the equilibrium will shift to counteract any stress that is placed on it. If the temperature is decreased, the equilibrium will shift toward the side that produces energy or heat. That's the product side where I_2 is produced.

27. **C** Because the equation is balanced, the following will occur:

If 1 mole of CH_3OH was created, then 1 mole of CO was consumed and 1 mole of CO remains; and if 1 mole of CH_3OH was created, then 2 moles of H_2 were consumed and 1 mole of H_2 remains. So at equilibrium, there are

(1 mol CH_3OH) + (1 mol CO) + (1 mol H_2) = 3 moles of gas

28. **D** According to Le Châtelier's law, equilibrium will shift to relieve any stress placed on a system. If the temperature is increased, the equilibrium will shift to favor the endothermic reaction because it absorbs the added energy. In this case, the equilibrium will be shifted to the right, increasing the concentration of both K^+ and NO_3^- ions.

Changing the volume of the solution, (A) and (B), will change the *number* of K^+ ions in solution, but not the *concentration* of K^+ ions. Because solids are not considered in the equilibrium expression, adding more solid KNO_3 to the solution (C) will not change the equilibrium. Decreasing the temperature (E) will favor the exothermic reaction, driving the equilibrium toward the left and decreasing the concentration of K^+ ions.

29. **D** K_p and K_c will differ the most in situations where the moles of gas change drastically from the reactants to the products. This relation is given in the equation

$$K_p = K_c(RT)^{\Delta n}$$

The exponential Δn is the moles of gas in the products minus the moles of gas in the reactants. The biggest change in number of moles of gas is found in the equation in choice D.

30. **D** The information about K at 200°C is useless information and can be ignored. Since all the Cl_2 found in solution must have come from $SbCl_5$, we know that at equilibrium $[Cl_2]$ = $[SbCl_3]$ = 0.1 M, and $[SbCl_5]$ = (1.0 − 0.1) M = .99 M. We can then say that K = (0.1)(0.1)/.99 = .0101 which is most closely approximated by choice D.

31. **D** $K_p = \dfrac{[NO]^2[Br_2]}{[NOBr]^2} = \dfrac{(4)^2(2)}{(4)^2} = 2$

32. **A** The equilibrium expression for the reaction is as follows:

$$\frac{P_{IBr}^{\ 2}}{P_{Br_2}P_{I_2}} = 300$$

When all of the values are plugged into the expression, (A) is the only choice that works.

$$\frac{(3)^2}{(0.1)(0.3)} = \frac{9}{0.03} = 300$$

33. **D** Use a table to see how the partial pressures change. Based on the balanced equation, we know that if 2 atm of $CO(g)$ were formed, then 2 atm of $H_2O(g)$ must also have formed. We also know that the reactants must have lost 2 atm each.

	$H_2(g)$	$CO_2(g)$	$H_2O(g)$	$CO(g)$
Before	6 atm	4 atm	0	0
Change	−2	−2	+2	+2
At Equilibrium	4 atm	2 atm	2 atm	2 atm

Now plug the numbers into the equilibrium expression.

$$K_{eq} = \frac{P_{H_2O}P_{CO}}{P_{H_2}P_{CO_2}} = \frac{(2)(2)}{(4)(2)} = \frac{1}{2}$$

34. **B** Every mole of $Mg(OH)_2$ molecules dissociates to produce 2 moles of OH^- ions, so a 0.10 M $Mg(OH)_2$ solution will be a 0.20 M OH^- solution.

The solution will be neutralized when the number of moles of H^+ ions added is equal to the number of OH^- ions originally in the solution.

Moles = (molarity)(volume)

Moles of OH^- = (0.20 M)(50 ml) = 10 millimoles = moles of H^+ added

$$Volume = \frac{moles}{molarity}$$

$$Volume\ of\ HCl = \frac{(10\,millimoles)}{(0.05\ M)} = 200\ ml$$

35. **B** You can eliminate (C), (D), and (E) by using common sense. HBrO is a weak acid, so an HBrO solution will be acidic, with a pH of less than 7.

To choose between (A) and (B) you have to remember that HBrO is a weak acid. If HBrO were a strong acid, it would dissociate completely and $[H^+]$ would be equal to 0.01-molar, for a pH of exactly 2. Because HBrO is a weak acid, it will not dissociate completely and $[H^+]$ will be less than 0.01-molar, which means that the pH will be greater than 2. So by POE, (B) is the answer.

36. **D** To find the lowest pH, we should look for the salt that produces an acidic solution.

The salt composed of the conjugate of a weak base (NH_4^+ is the conjugate of NH_3) and the conjugate of a strong acid (Cl^- is the conjugate of HCl) will produce an acidic solution.

As for the other choices, (A) and (C) are composed of conjugates of strong acids and bases and will produce neutral solutions.

(B) is composed of the conjugate of a strong base (Na^+) and the conjugate of a weak acid ($C_2H_3O_2^-$) and will produce a basic solution.

37. **A** A pH of greater than 7 means that the solution is basic.

The salt composed of the conjugate of a strong base (Na^+ is the conjugate of NaOH) and the conjugate of a weak acid (CN^- is the conjugate of HCN) will produce a basic solution.

As for the other choices, (B) and (C) are composed of conjugates of strong acids and bases and will produce neutral solutions.

(D) is composed of the conjugate of a weak base (NH_4^+) and the conjugate of a strong acid (NO_3^-) and will produce an acidic solution.

38. **C** The best buffered solution occurs when pH = pK_a. That happens when the solution contains equal amounts of acid and conjugate base. If you want to create a buffer with a pH of 5, the best choice would be an acid with a pK_a that is as close to 5 as possible. You shouldn't have to do a calculation to see that the pK_a for choice (C) is much closer to 5 than that of any of the others.

39. **C** An amphoteric species can act either as an acid or a base, gaining or losing a proton.

HCO_3^- can act as an acid, losing a proton to become CO_3^{2-}, or it can act as a base, gaining a proton to become H_2CO_3.

40. **D** We want to change the hydrogen ion concentration from 0.1 M (pH of 1) to 0.01 M (pH of 2).

The HCl is completely dissociated, so the number of moles of H^+ will remain constant as we dilute the solution.

Moles = (molarity)(volume) = Constant

$(M_1)(V_1) = (M_2)(V_2)$

$(0.1\ M)(1\ L) = (0.01\ M)(V_2)$

So, V_2 = 10 L, which means that 9 L must be added.

41. **A** A pH of 5 means that $[H^+] = 1 \times 10^{-5}$

$$K_a = \frac{[H^+][A^-]}{[HA]}$$

For every HA that dissociates, we get one H^+ and one A^-, so $[H^+] = [A^-] = 1 \times 10^{-5}$.

The acid is weak, so we can assume that very little HA dissociates and that the concentration of HA remains 1-molar.

$$K_a = \frac{[H^+][A^-]}{[HA]} = \frac{(1 \times 10^{-5})(1 \times 10^{-5})}{(1)} = 1 \times 10^{-10}.$$

42. **A** For conjugates, $(K_a)(K_b) = K_w = 1 \times 10^{-14}$

$$K_b = \frac{K_w}{K_a} = \frac{(1 \times 10^{-14})}{(1 \times 10^{-2})} = 1 \times 10^{-12}$$

43. **D** First let's find out how many moles of H^+ ions we need to neutralize.

Every H_2SO_3 will produce 2 H^+ ions, so for our purposes, we can think of the solution as a 0.4-molar H^+ solution.

Moles = (molarity)(volume)

Moles of H^+ = (molarity)(volume) = (0.4 M)(100 ml) = 40 millimoles = moles of OH^- required

$$\text{Volume of NaOH} = \frac{\text{moles}}{\text{molarity}} = \frac{(40 \text{ millimoles})}{(0.1 \, M)} = 400 \text{ ml}$$

44. **C** An amphoteric species can act either as a base and gain an H^+ or as an acid and lose an H^+.

HSO_4^- can lose an H^+ to become SO_4^{2-} or it can gain an H^+ to become H_2SO_4.

HNO_3 (A), $HC_2H_3O_2$ (B), and H_3PO_4 (D) can lose only an H^+.

45. **D** Every unit of HNO_3 added to the solution will place 1 unit of H^+ ions in the solution. So first find the moles of HNO_3 added.

$$\text{Moles} = \frac{\text{grams}}{\text{MW}}$$

$$\text{Moles of } H^+ = \frac{0.630 \text{ grams}}{63.0 \text{ g / mole}} = 0.01 \text{ moles}$$

Now it's easy to find the H^+ concentration.

$$\text{Molarity} = \frac{\text{moles}}{\text{liters}}$$

$$[H^+] = \frac{0.01 \text{ moles}}{1 \text{ L}} = 0.01 \ M$$

$$pH = -\log [H^+] = -\log(0.01) = 2$$

46. **D** A buffered solution can be prepared by mixing a weak acid with an equal amount of its conjugate or by adding enough strong base to neutralize half of the weak acid present in a solution.

In (I), equal amounts of a strong acid and base are mixed. They'll neutralize each other completely to produce salt water, which is not a buffer.

In (II), enough strong base is added to neutralize half of the weak acid. This will leave equal amounts of weak acid and its conjugate base, producing a buffered solution.

In (III), equal amounts of a weak acid and its conjugate base are mixed. This will produce a buffered solution.

47. **A** A Lewis acid must be able to accept a pair of electrons. By drawing the Lewis structure of $AlBr_3$, shown below, one can see that Al lacks a full octet and therefore is a very strong Lewis acid.

NH_3 is a good Lewis base, since it has a lone pair of electrons to donate. Both CH_4 and Xe have full octets, and will not accept more electrons. The compound PCl_3 looks a lot like NH_3, and has a lone pair of electrons and acts as a Lewis base.

48.   **D**   The half equivalence point of this titration, shown in the plot below, is around pH = 9.8.

Mol acid titrated

However, this is not the pK$_b$ of the weak base. Remember that the Henderson-Hasselbalch equation relates pH with pK$_a$, not pK$_b$. At the half equivalence point shown, the amount of base (which we can call [A$^-$] as we do when talking about acids and pH values) is equal to the neutralized base [HA]. When these two quantities are equal, the H-H equation tells us that the pH = the pK$_a$ of the acid form, which when subtracted from 14 gives the pK$_b$ of the base. So, the pK$_a$ of the acid form is 9.8, so the pK$_b$ of the base is 4.2.

Free Response Questions

1. (a) The solubility product is the same as the equilibrium expression, but because the reactant is a solid, there is no denominator.

$$K_{sp} = [Ca^{2+}][OH^-]^2$$

(b) Use the solubility product.

$$K_{sp} = [Ca^{2+}][OH^-]^2$$

$$5.5 \times 10^{-6} = (x)(2x)^2 = 4x^3$$

$$x = 0.01 \ M \text{ for } Ca^{2+}$$

One mole of calcium hydroxide produces 1 mole of Ca^{2+}, so the concentration of $Ca(OH)_2$ must be 0.01 M.

Moles = (molarity)(volume)

Moles of $Ca(OH)_2$ = (0.01 M)(0.500 L) = 0.005 moles

Grams = (moles)(MW)

Grams of $Ca(OH)_2$ = (0.005 mol)(74 g/mol) = 0.37 g

(c) We can find $[OH^-]$ from (b).

If $[Ca^{2+}] = 0.01 \ M$, then $[OH^-]$ must be twice that, so $[OH^-] = 0.02 \ M$

$$pOH = -\log[OH^-] = 1.7$$

$$pH = 14 - pOH = 14 - 1.7 = 12.3$$

(d) Find the new $[OH^-]$. The hydroxide already present is small enough to ignore, so we'll use only the hydroxide just added.

$$\text{Molarity} = \frac{\text{moles}}{\text{liters}}$$

$$[OH^-] = \frac{(1.0 \text{ mol})}{(0.500 \text{ L})} = 2.0 \ M$$

Now use the K_{sp} expression.

$$K_{sp} = [Ca^{2+}][OH^-]^2$$

$$5.5 \times 10^{-6} = [Ca^{2+}](2.0 \ M)^2$$

$$[Ca^{2+}] = 1.4 \times 10^{-6} \ M$$

2. (a) Entropy increases when a salt dissociates because aqueous particles have more randomness than a solid.

(b) Most salt solution processes are endothermic, and endothermic processes are favored by an increase in temperature, therefore increasing temperature will increase the solubility of most salts.

(c) $Ce_2(SO_4)_3$ becomes less soluble as temperature increases, so the solution process for this salt must be exothermic.

HCl is a strong acid, which means that it dissociates completely. This means that 1 mole of HCl in solution will produce 2 moles of particles. HF is a weak acid, which means that it dissociates very little. This means that 1 mole of HF in solution will remain at about 1 mole of particles in solution.

Therefore, the HCl solution will have more particles than the HF solution.

3. (a) Use the equilibrium expression.

$$K_1 = \frac{\left[H^+\right]\left[HCO_3^-\right]}{\left[H_2CO_3\right]}$$
$$\left[H^+\right] = \left[HCO_3^-\right] = x$$
$$\left[H_2CO_3\right] = (0.050\,M - x)$$

Assume that x is small enough that we can use $[H_2CO_3]$ = (0.050 M).

$$4.3 \times 10^{-7} = \frac{x^2}{(0.050)}$$
$$x = \left[H^+\right] = 1.5 \times 10^{-4}$$
$$pH = -\log\left[H^+\right] = -\log\left(1.5 \times 10^{-4}\right) = 3.8$$

(b) Use the equilibrium expression.

$$K_2 = \frac{\left[H^+\right]\left[CO_3^{2-}\right]}{\left[HCO_3^-\right]}$$

From (a) we know that $[H^+] = \left[HCO_3^-\right] = 1.5 \times 10^{-4}$.

$$5.6 \times 10^{-11} = \frac{\left(1.5 \times 10^{-4}\right)\left[CO_3^{2-}\right]}{\left(1.5 \times 10^{-4}\right)} = \left[CO_3^{2-}\right]$$
$$\left[CO_3^{2-}\right] = 5.6 \times 10^{-11}\,M$$

(c)   (i)   Adding HCl will increase [H⁺], lowering the pH.

     (ii)   From Le Châtelier's law, you can see that adding $NaHCO_3$ will cause the first equilibrium to shift to the left to try to use up the excess HCO_3^-. This will cause a decrease in [H⁺], raising the pH.

        You may notice that adding $NaHCO_3$ will also cause the second equilibrium to shift toward the right, which should increase [H⁺], but because K_2 is much smaller than K_1, this shift is insignificant.

     (iii)   Adding NaOH will neutralize hydrogen ions, decreasing [H⁺] and raising the pH.

     (iv)   Adding NaCl will have no effect on the pH.

(d)   The reaction in (d) is just the sum of the two reactions given. When two reactions can be added to give a third reaction, the equilibrium constants for those reactions can be multiplied to give K_{eq} for the third reaction.

$$K_{eq} = (K_1)(K_2) = (4.3 \times 10^{-7})(5.6 \times 10^{-11}) = 2.4 \times 10^{-17}$$

4.   (a) $K_c = \dfrac{[NH_3]^2}{[N_2][H_2]^3}$

$$K_c = \dfrac{(0.20)^2}{(0.50)(0.20)^3} = 10$$

(b)   Use the formula that relates the two constants.

$$K_p = K_c(RT)^{\Delta n}$$

Δn is the change in the number of moles of gas from reactants to products. So $\Delta n = -2$.

$$K_p = (10)\left[(0.082)(570)\right]^{-2} = (10)(46.7)^{-2} = 4.7 \times 10^{-3}$$

(c) (i) An increase in temperature favors the endothermic direction. In this case, that's the reverse reaction, so the concentration of H_2 will increase.

(ii) An increase in volume favors the direction that produces more moles of gas. In this case, that's the reverse direction, so the concentration of H_2 will increase.

(iii) According to Le Châtelier's law, increasing the concentration of the reactants forces the reaction to proceed in the direction that will use up the added reactants. In this case, adding the reactant N_2 will shift the reaction to the right and decrease the concentration of H_2.

(iv) The addition of He, a gas that takes no part in the reaction, will have no effect on the concentration of H_2.

5. (a) Use the equilibrium expression for H_2S to find $[H^+]$.

$$K_1 = \frac{[H^+][HS^-]}{[H_2S]}$$

$[H^+] = [HS^-] = x$

$[H_2S] = 0.100\ M$

$K_1 = 1.0 \times 10^{-7}$

$$1.0 \times 10^{-7} = \frac{x^2}{(0.100\ M - x)}$$

Let's assume that 0.100 is much larger than x. That simplifies the expression.

$$1.0 \times 10^{-7} = \frac{x^2}{(0.100\ M)}$$

$x^2 = (1.0 \times 10^{-7})(0.100)\ M^2 = 1.0 \times 10^{-8}\ M^2$

$x = 1.0 \times 10^{-4}\ M = [H^+]$

$pH = -\log[H^+] = -\log(1.0 \times 10^{-4}) = 4.0$

(b) At the midpoint, pH = pK_1

Let's look at the Henderson-Hasselbalch expression.

$$pH = pK + \log \frac{[HS^-]}{[H_2S]}$$

When pH = pK, $[HS^-]$ must be equal to $[H_2S]$, making $\frac{[HS^-]}{[H_2S]}$ equal to one and $\log \frac{[HS^-]}{[H_2S]}$ equal to zero.

For every molecule of H_2S neutralized, one unit of HS^- is generated, so we must add enough NaOH to neutralize half of the HS^- initially present. (We are assuming that the further dissociation of HS^- into H^+ and S_2^- is negligible and can be ignored.)

Let's find out how many moles of H_2S were initially present.

Moles = (molarity)(volume)

Moles of H_2S = (0.100 M)(0.500 L) = 0.050 moles

We need to neutralize half of that, or 0.025 moles, so we need 0.025 moles of NaOH.

$$Volume = \frac{moles}{molarity}$$

Volume of NaOH = $\frac{(0.025 \text{ mol})}{(0.100 \text{ } M)}$ = 0.250 L = 250 ml

(c) The concentrations of the H_2S solution and NaOH solution are both 0.100 M, so the first 500 ml of NaOH solution will completely neutralize the H_2S. The final 300 ml of NaOH solution will neutralize HS^- molecules.

Initially, there were 0.050 moles of H_2S, so after 500 ml of NaOH solution was added, there were 0.050 moles of HS^-.

By adding 300 ml more of NaOH solution, we added (0.100 M) (0.300 L) = 0.030 moles of NaOH. [Moles = (molarity)(volume)]

Every unit of NaOH added neutralizes one unit of HS^-, so when all the NaOH has been added, we have 0.030 moles of S^{2-} and (0.050 moles) − (0.030 moles) = 0.020 moles of HS^-.

Now we can use the Henderson-Hasselbalch expression to find the pH. We can use the number of moles we just calculated (0.030 for S^{2-} and 0.020 for HS^-) instead of concentrations because in a solution the concentrations are proportional to the number of moles.

$$pH = pK + \log \frac{\left[S^{2-}\right]}{\left[HS^-\right]}$$

$$pK = -\log(1.3 \times 10^{-13}) = 12.9$$

$$\log \frac{\left[S^{2-}\right]}{\left[HS^-\right]} = \log \frac{(0.030)}{(0.020)} = \log(1.5) = 0.18$$

$$pH = 12.9 + 0.18 = 13.1.$$

(d) The reaction ($H_2S \rightleftharpoons 2\,H^+ + S^{2-}$) is the sum of the two acid dissociation reactions below.

$$H_2S \rightleftharpoons H^+ + HS^-$$

$$HS^- \rightleftharpoons H^+ + S^{2-}$$

If one reaction is the sum of two other reactions, then its equilibrium constant will be the product of the equilibrium constants for the other two reactions.

So $K_{eq} = K_1 K_2 = (1.0 \times 10^{-7})(1.3 \times 10^{-13}) = 1.3 \times 10^{-20}$

6. (a) $pK_b = -\log K_b = -\log(1.79 \times 10^{-5}) = 4.75$

(b) This is a buffered solution, so we'll use the Henderson-Hasselbalch expression.

First let's find $[NH_4^+]$ and $[NH_3]$.

Moles = (molarity)(volume)

Moles of NH_4^+ = (0.100 M)(0.100 L) = 0.010 moles

Moles of NH_3 = (0.200)(0.080 L) = 0.016 moles

When we mix the solutions, the volume becomes (0.100 L) + (0.080 L) = 0.180 L.

$$Molarity = \frac{moles}{volume}$$

$$[NH_4^+] = \frac{(0.010\ mol)}{(0.180\ L)} = 0.056\ M$$

$$[NH_3] = \frac{(0.016 \text{ mol})}{(0.180 \text{ mol})} = 0.089 \text{ } M$$

Now we can use the Henderson-Hasselbalch expression for bases.

$$pOH = pK + \log \frac{[NH_4^+]}{[NH_3]}$$

$$pK = 4.75$$

$$\log \frac{[NH_4^+]}{[NH_3]} = \log \frac{(0.056 \text{ } M)}{(0.089 \text{ } M)} = -0.20$$

$$pOH = 4.75 + (-0.20) = 4.55$$

$$pH = 14 - pOH = 14 - 4.55 = 9.45$$

(c) First let's find out how many moles of NaOH were added.

$$\text{Moles} = \frac{\text{moles}}{\text{MW}}$$

$$\text{Moles of NaOH} = \frac{(0.200 \text{ g})}{(40.0 \text{ g/m})} = 0.005 \text{ mol}$$

When NaOH is added to the solution, the following reaction occurs:

$$NH_4^+ + OH^- \rightarrow NH_3 + H_2O$$

So for every unit of NaOH added, one ion of NH_4^+ disappears and one molecule of NH_3 appears. We can use the results of the molar calculations we did in part (b).

$$\text{Moles of } NH_4^+ = (0.010) - (0.005) = 0.005 \text{ moles}$$

$$\text{Moles of } NH_3 = (0.016) + (0.005) = 0.021 \text{ moles}$$

Now we can use the Henderson-Hasselbalch expression. We can use the number of moles we just calculated (0.005 for NH_4^+ and 0.021 for NH_3) instead of concentrations because in a solution the concentrations will be proportional to the number of moles.

$$pOH = pK + \log \frac{[NH_4^+]}{[NH_3]}$$

$$pK = 4.75$$

$$\log \frac{[NH_4^+]}{[NH_3]} = \log \frac{(0.005 \text{ mol})}{(0.021 \text{ mol})} = -0.62$$

$$pOH = 4.75 + (-0.62) = 4.13$$

$$pH = 14 - pOH = 14 - 4.13 = 9.87$$

(d) When equal quantities of a base (NH_3) and its conjugate acid (NH_4^+) are mixed in a solution, the pOH will be equal to the pK_b.

From (a), $pOH = pK_b = 4.75$

$pH = 14 - pOH = 14 - 4.75 = 9.25$.

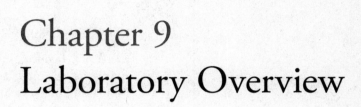

Chapter 9
Laboratory Overview

DISCLAIMER

Please note that at the time of this book's printing, the new laboratory exercises have not been made available. There are now 16 laboratory exercises that the AP Chemistry exam could potentially cover. For more information about these laboratory exercises, please visit http://advancesinap.collegeboard.org/science/chemistry and scroll down to the AP Chemistry Labs section. In this section, we will discuss safety and accuracy precautions, laboratory equipment, and laboratory procedures.

SAFETY

Here are some basic safety rules that might turn up in test questions.

- Don't put chemicals in your mouth. You were told this when you were four years old, and it still holds true for the AP Chemistry Exam.
- When diluting an acid, always add the acid to the water. This is to avoid the spattering of hot solution.
- Always work with good ventilation; many common chemicals are toxic.
- When heating substances, do it slowly. When you heat things too quickly, they can spatter, burn, or explode.

ACCURACY

Here are some rules for ensuring the accuracy of experimental results.

- When titrating, rinse the buret with the solution to be used in the titration instead of with water. If you rinse the buret with water, you might dilute the solution, which will cause the volume added from the buret to be too large.
- Allow hot objects to return to room temperature before weighing. Hot objects on a scale create convection currents that may make the object seem lighter than it is.
- Don't weigh reagents directly on a scale. Use a glass or porcelain zcon- tainer to prevent corrosion of the balance pan.
- When collecting a gas over water, remember to take into account the pressure and volume of the water vapor.
- Don't contaminate your chemicals. Never insert another piece of equipment into a bottle containing a chemical. Instead you should always pour the chemical into another clean container. Also, don't let the inside of the stopper for a bottle containing a chemical touch another surface.
- When mixing chemicals, stir slowly to ensure even distribution.
- Be conscious of significant figures when you record your results. The number of significant figures you use should indicate the accuracy of your results.
- Be aware of the difference between accuracy and precision. A measurement is accurate if it is close to the accepted value. A series of measurements is precise if the values of all of the measurements are close together.

LAB PROCEDURES

METHODS OF SEPARATION

Filtration—In filtration, solids are separated from liquids when the mixture is passed through a filter. Typically, porous paper is used as the filter. To find the amount of solid that is filtered out of a mixture, the filter paper containing the solid is allowed to dry and then weighed. The initial weight of the clean, dry filter is then subtracted from the weight of the dried filter paper and solid.

Distillation—In distillation, the differences in the boiling points of liquids are used to separate them. The temperature of the mixture is raised to a temperature that is greater than the boiling point of the more volatile substance and lower than the boiling point of the less volatile substance. The more volatile substance will vaporize, leaving the less volatile substance.

Chromatography—In chromatography, substances are separated by the differences in the degree to which they are adsorbed onto a surface. The substances are passed over the adsorbing surface, and the ones that stick to the surface with greater attraction will move slower than the substances that are less attracted to the surface. This difference in speeds separates the substances. The name chromatography came about because the process is used to separate pigments.

MEASURING CONCENTRATION

Titration is one of the most important laboratory procedures. In titration, an acid-base neutralization reaction is used to find the concentration of an unknown acid or base. It takes exactly one mole of hydroxide ions (base) to neutralize one mole of hydrogen ions (acid), so the concentration of an unknown acid solution found by calculating how much of a known basic solution is required to neutralize a sample of given volume. The most important formula in titration experiments is derived from the definition of molarity.

Molarity = Moles/Liters
Moles = (Molarity)(Liters)

The moment when exactly enough base has been added to the sample to neutralize the acid present is called the equivalence point. In the lab, an indicator is used to tell when the equivalence point has been reached. An indicator is a substance that is one color in acid solution and a different color in basic solution. Two popular indicators are phenolphthalein, which is clear in acidic solution and pink in basic solution, and litmus, which is pink in acidic solution and blue in basic solution.

Spectrophotometer—A spectrophotometer measures slight variations in color. It can be used to measure the concentration of ions that produce colored solutions.

The concentration of the ions will be directly proportional to the absorbance of the solution measured by the spectrophotometer. The relationship between absorbance and concentration is given by **Beer's law.**

> ### Beer's Law
> $$A = \varepsilon l c$$
>
> A = absorbance measured by spectrophotometer
> ε = molar absorptivity, a constant whose value depends on the solution and the wavelength of light used
> l = path length, the distance the light travels through the solution
> c = concentration of the solution (mol/L)

IDENTIFYING CHEMICALS IN SOLUTION

Precipitation

One of the most useful ways of identifying unknown ions in solution is precipitation. You can use the solubility rules given in the solubility chapter to see how the addition of certain ions to solution will cause the specific precipitation of other ions. For instance, the fact that $BaSO_4$ is insoluble can be used to identify either can be used to identify either Ba^{2+} or SO_4^{2-} in solution; if the solution contains Ba^{2+} ions, then the addition of SO_4^{2-} will cause a precipitation reaction. The inverse is true for a solution that contains SO_4^{2-} ions.

In the same way, the insolubility of AgCl can be used to identify either Ag^+ or Cl^- in solution.

Conduction

You can tell whether a solution contains ions by checking to see if the solution conducts electricity. Ionic solutes conduct electricity in solution; nonionic solutes do not.

Flame Tests

Some ions burn with distinctly colored flames. Flame tests can be used to identify Li^+ (red), Na^+ (yellow), and K^+ (purple), as well as the other alkali metals. The alkaline earths, including Ba^{2+} (green), Sr^{2+} (red), and Ca^{2+} (red) also burn with colored flames.

Acid-Base Reaction

- When a base is added to an NH_4^+ solution, the distinctive odor of ammonia can be detected.
- When an acid is added to a solution containing S_2^-, the rotten-egg odor of H_2S can be detected.
- When acid is added to a solution containing CO_3, CO_2 gas is produced.

LABORATORY EQUIPMENT

The pictures below show some standard chemistry lab equipment.

Beaker	Used to hold and pour liquids— also your favorite muppet.
Buret	Used to add small but precisely measured volumes of liquid to a solution, used frequently in titration experiments.
Burner	Used to apply heat and to wake up sleeping AP Chem students.

Crucible tongs	Used to handle objects that are too hot to touch (careful though, they can break test tubes!).
Dropper pipette	Used to add small amounts of liquid to a solution, but only when a precise volume is not needed because the drops themselves should be consistent only for one particular dropper (i.e. adding an indicator, comparing the amount of drops of strong acid or base needed for a pH change in different solutions).
Erlenmeyer flask	A flask used for heating liquids. The conic shape allows stirring.
Evaporating dish	Used to hold liquids for evaporation. The wide mouth allows vapor to escape.
Florence flask	Used for boiling of liquids. The small neck prevents excessive evaporation and splashing.
Forceps	A fancy name for tweezers.

Funnel	Used to get liquids into a smaller container. Doubles as a cute hat when inverted.
Graduated cylinder	Used for measuring precisely a volume of liquid to be poured all at once (rather than dripped from a buret).
Graduated pipette	Used to transfer small and precise volumes of liquid from one container to another (gradations indicate the volume).
Metal spatula	Used to scoop and transport powders.
Mortar and pestle	Used to grind solids into powders suitable for dissolving or mixing.
Pipette bulb	Rubber bulb used to draw liquid into pipette. In a pinch could be used as a miniature clown nose.

Platform balance (triple beam)	A very precise scale operated by moving a set of three weights (typically corresponding to 100, 10, and 1 gram increments). A measurement will proceed like this: Rear weight is in the notch reading 30 g Middle weight is in the notch reading 200 g Front beam weight reads 3.86 g The sample weighs 200 + 30 + 3.86 = 233.86 g
Ring clamp	Used to hold funnels or other vessels in conjunction with a stand.
Rubber policeman	A hard tipped rubber scraper used to transfer precipitate.
Safety goggles	Used by chemists worldwide to protect their eyes during all laboratory experiments.
Test tube	Used to contain samples, especially when heating. Sometimes babies come out of them.

Thermometer	Measures temperature of a solution (don't use these to measure your body temperature, you don't know where they've been).
Volumetric pipette	Used to transfer small amounts of liquid. The big difference between a graduated pipette and a volumetric pipette is that the volumetric type is suited for only one particular volume. Because of this, they are extremely accurate.

CHAPTER 9 QUESTIONS

Multiple-Choice Questions

Questions 1–4

    (A) Oxidation-reduction
    (B) Neutralization
    (C) Fusion
    (D) Combination

Which of the reaction types listed above best describes each of these processes?

1. $CO_2(g) + CaO(s) \rightarrow CaCO_3(s)$

2. $2\ Fe^{3+}(aq) + 2\ I^-(aq) \rightarrow 2\ Fe^{2+}(aq) + I_2(aq)$

3. $CH_3COOH(aq) + NaOH(aq) \rightarrow CH_3COONa(aq) + H_2O(l)$

4. $CH_4(g) + 2\ O_2(g) \rightarrow CO_2(g) + 2\ H_2O(g)$

Questions 5–7

    (A) Na^+
    (B) Cu^{2+}
    (C) Ag^+
    (D) Al^{3+}

5. This ion turns an aqueous solution deep blue.

6. This ion forms a white precipitate when added to a solution containing chloride ions.

7. This ion produces a yellow flame when burned.

8. Which of the following indicators would be most useful in identifying the equivalence point of a titration for a solution that has a hydrogen ion concentration of 7×10^{-4} M at the equivalence point?

(A) Methyl violet (pH range for color change is 0.1–2.0)
(B) Methyl yellow (pH range for color change is 1.2–2.3)
(C) Methyl orange (pH range for color change is 2.9–4.0)
(D) Methyl red (pH range for color change is 4.3–6.2)

9. The volume of a liquid is to be measured. Which of the following cylindrical flasks would take the most accurate measurement?

(A) A flask with 1 ml gradations and a diameter of 1 cm
(B) A flask with 1 ml gradations and a diameter of 3 cm
(C) A flask with 5 ml gradations and a diameter of 1 cm
(D) A flask with 5 ml gradations and a diameter of 3 cm

10. Which of the following is (are) considered to be proper laboratory procedure?

   I. Reading the height of a fluid in a buret from a point level with the fluid's meniscus.
   II. Placing a sample to be weighed directly on the pan of the balance.
   III. Stirring a solution constantly during a titration.

(A) I only
(B) III only
(C) I and III only
(D) I, II, and III

11. A 0.1-molar NaOH solution is to be released from a buret in a titration experiment to measure the hydrogen ion concentration of an unknown acid. Which of the following laboratory procedures would cause an error in the measure of the concentration of the acid?

   I. The buret was rinsed with the NaOH solution immediately before the titration.
   II. The buret was rinsed with distilled water immediately before the titration.
   III. The buret was rinsed with the unknown acid immediately before the titration.

(A) I only
(B) III only
(C) I and II only
(D) II and III only

12. Which of the following is NOT proper procedure for transferring a solution with a pipette?

(A) Rinsing the pipette with the solution to be transferred.
(B) Using your mouth to draw the solution into the pipette.
(C) Covering the top of the pipette with your index finger to keep the solution from escaping.
(D) Draining the solution into a waste beaker until the meniscus drops to the calibration mark.

13. An object that was weighed on a balance was later found to be slightly heavier than the weight that was recorded by the balance. Which of the following could have caused the discrepancy?

    I.   There was some foreign matter on the weighing paper along with the object.
   II. The object was hot when it was weighed.
  III. The experimenter neglected to account for the weight of the weighing paper.

(A) I only
(B) II only
(C) I and II only
(D) I, II, and III

14. An experimenter wishes to use test paper to find the pH of a solution. Which of the following is part of the proper procedure for this process?

(A) Dipping the test paper in the solution while stirring.
(B) Dipping the test paper in the solution without stirring.
(C) Pouring some of the solution onto the dry test paper.
(D) Dipping the test paper in distilled water and slowly adding the solution to the water while stirring.

Free Response Questions

1. An experiment was conducted to determine the molecular weight of a pure salt sample. The mass of the salt sample was known. The salt was dissolved in a container of water of known mass and the freezing point of the solution was measured. The molecular weight was calculated by the freezing point depression method. How would the calculated value of the molecular weight be affected by each of the following?

   (a) The experimenter failed to take the dissociation of the salt into account.
   (b) The experimenter mistook molarity for molality, and used liters of solution instead of kilograms of solvent in the calculation to find the number of moles of solute.
   (c) The container used for the experiment was not rinsed and contained dust particles.
   (d) The experimenter misread the thermometer and recorded a freezing point that was higher than the true value.
   (e) The experimenter did not notice that some solid salt did not completely dissolve.

2. Use your knowledge of chemical principles to answer or explain each of the following:

   (a) When helium gas is to be collected in a jar by the displacement of air, the opening of the jar must be directed downward. When carbon dioxide gas is to be collected in a jar by the displacement of air, the opening of the jar must be directed upward.
   (b) Will the molar quantity calculated for a gas collected over water be too large or too small if the experimenter fails to take into account the vapor pressure of water?
   (c) Why is it easier to separate oxygen gas from hydrogen gas by the method of successive effusion than it is to separate oxygen gas from nitrogen gas by the same method?
   (d) Give an explanation for why an attempt to separate two liquids by distillation may fail.

CHAPTER 9 ANSWERS AND EXPLANATIONS

Multiple-Choice Questions

1. **D** In a combination (or composition, or synthesis) reaction, two substances combine to form a more complex substance.

   A combination reaction is the opposite of a decomposition reaction, so if this reaction occurred in reverse, it would be a decomposition reaction.

2. **A** In an oxidation-reduction reaction, electrons are transferred between the reactants, causing the oxidation state of the element that is oxidized to increase and the oxidation state of the element that is reduced to decrease. In this reaction, Fe^{3+} is reduced to Fe^{2+} and I^- is oxidized to I^0.

3. **B** In a neutralization reaction, an acid (in this case, CH_3COOH) and a base ($NaOH$) react to form water and a salt (CH_3COONa).

4. **A** This reaction, the combustion of an organic compound, is also an oxidation-reduction reaction. As we said above, in a redox reaction, electrons are transferred between the reactants, causing the oxidation state of the element that is oxidized to increase and the oxidation state of the element that is reduced to decrease. In this reaction, H^- is oxidized to H^+ and O^0 is reduced to O^{2-}.

5. **B** Copper, along with most of the transition metals, forms colored solutions with water. This is true because the d electrons of the transition metals are constantly changing energy levels and emitting radiation in the visible spectrum.

6. **C** Silver is the only ion listed that forms an insoluble chloride.

7. **A** Sodium, along with the other Group IA elements, produces a colored flame in the flame test.

8. **C** If $[H^+]$ is 7×10^{-4} M, then the pH must be between 3 and 4. Only methyl orange changes color between 3 and 4.

9. **A** Volume = (height)(cross-sectional area)

   The smaller the gradations, the more accurately the height can be measured. The smaller the area, the farther apart the 1 ml gradations will be and the more accurately the height of the fluid can be measured.

10.　**C**　Choices (I) and (III) are proper experimental procedures.

(II) is not; a sample should always be weighed in a glass or porcelain container to prevent a reaction with the balance pan.

11.　**D**　Rinsing the buret with the NaOH solution (I) is proper procedure and will not change the concentration of the NaOH solution and will not cause an error. Rinsing the buret with distilled water (II) will dilute the NaOH solution, lowering the concentration and causing an error, and rinsing the buret with the unknown acid (III) will cause some of the NaOH solution to be neutralized in the buret, lowering its concentration and causing an error.

12.　**B**　You should never use your mouth to draw solution into a pipette. Instead, you should use a rubber suction bulb.

All of the other choices are part of the proper procedure.

13.　**B**　When a hot object is weighed (II), convection currents around the object can reduce the apparent mass measured by the balance.

Choices (I) and (III) would both cause the weight measured by the balance to be greater than the actual weight of the object. We're looking for the opposite effect.

14.　**C**　(A) and (B) are wrong because there is a danger of contaminating the solution by adding the paper.

(D) is wrong because adding the solution to distilled water completely changes the solution and defeats the purpose of testing it.

So pouring the solution onto the dry test paper (C) is the proper procedure.

Free Response Questions

1. A note for the answers:

   In this experiment, the freezing point of the solution is measured, and from the freezing point the freezing-point depression, ΔT, is calculated.

   The equation $m = \dfrac{T}{kx}$ is used to calculate the molality of the solution.

   Moles = (molality)(kg of solvent) is used to calculate the number of moles of salt.

   MW = $\dfrac{\text{moles}}{\text{grams}}$ is used to calculate the molecular weight of the salt.

   (a) The experimenter makes $x = 1$, instead of 2 or 3. So the calculated value of m will be too large, the calculated value of moles of salt will be too large, and the calculated MW will be too small.

   (b) Moles = (molality)(kg of solvent)
   Moles = (molarity)(liters of solution)

   Because the solvent is water, the distinction between kilograms and liters is not important (for water, 1kg = 1 L), but the distinction between solvent and solution may make a difference.

   Liters of solution will be a little larger than kilograms of solvent, so the calculated value for moles of salt will be too large, and the calculated MW will be too small.

   (c) Extra particles in the solution will cause the measured freezing point to be too low. ΔT will be too large, m will be too large, and the calculated MW will be too small.

   (d) If the freezing point is too high, the calculated ΔT will be too small, m will be too small, and the calculated MW will be too large.

   (e) If some salt does not dissolve, then the grams of salt used in the calculation will be larger than the amount actually in the solution, and the calculated MW will be too large.

2. (a) Helium (MW = 4 g/mol) is less dense than air, so it will rise to the top of the jar, displacing air downward.

Carbon dioxide (MW = 44 g/mol) is denser than air, so it will sink to the bottom of the jar, displacing air upward.

(b) If the vapor pressure from water is ignored, the pressure of the gas used in the calculation will be too large.

$$n = \frac{PV}{RT}$$

If P is too large, then n, the calculated molar quantity of gas, will be too large.

(c) Separation of gases by successive effusion depends on Graham's law.

$$\frac{v_1}{v_2} = \sqrt{\frac{MW_2}{MW_1}}$$

The greater the difference in molecular weights, the greater the difference in average molecular speeds, and the greater the difference in rates of effusion.

Oxygen gas (32 g/mol) and nitrogen gas (28 g/mol) have similar molecular weights.

Oxygen gas (32 g/mol) and hydrogen gas (2 g/mol) have very different molecular weights.

(d) The most likely reason for the failure of separation by distillation would be that the boiling points of the two liquids are too close together.

Part V
The Princeton Review AP Chemistry Practice Tests

Chapter 10
Practice Test 1

AP® Chemistry Exam

DO NOT OPEN THIS BOOKLET UNTIL YOU ARE TOLD TO DO SO.

At a Glance

Total Time
1 hour and 30 minutes
Number of Questions
60
Percent of Total Grade
50%
Writing Instrument
Pencil required

Instructions

Section I of this examination contains 60 multiple-choice questions. Fill in only the ovals for numbers 1 through 60 on your answer sheet.

CALCULATORS MAY NOT BE USED IN THIS PART OF THE EXAMINATION.

Indicate all of your answers to the multiple-choice questions on the answer sheet. No credit will be given for anything written in this exam booklet, but you may use the booklet for notes or scratch work. After you have decided which of the suggested answers is best, completely fill in the corresponding oval on the answer sheet. Give only one answer to each question. If you change an answer, be sure that the previous mark is erased completely. Here is a sample question and answer.

Sample Question   Sample Answer

Chicago is a
(A) state
(B) city
(C) country
(D) continent

Use your time effectively, working as quickly as you can without losing accuracy. Do not spend too much time on any one question. Go on to other questions and come back to the ones you have not answered if you have time. It is not expected that everyone will know the answers to all the multiple-choice questions.

About Guessing

Many candidates wonder whether or not to guess the answers to questions about which they are not certain. Multiple choice scores are based on the number of questions answered correctly. Points are not deducted for incorrect answers, and no points are awarded for unanswered questions. Because points are not deducted for incorrect answers, you are encouraged to answer all multiple-choice questions. On any questions you do not know the answer to, you should eliminate as many choices as you can, and then select the best answer among the remaining choices.

Disclaimer

This test is an approximation of the test that you will take and some of the question types may change. For up-to-date information, please remember to check the AP Central website.

CHEMISTRY
SECTION I

Time—1 hour and 30 minutes

Material in the following table may be useful in answering the questions in this section of the examination.

PERIODIC CHART OF THE ELEMENTS

1 H 1.0																	2 He 4.0
3 Li 6.9	4 Be 9.0											5 B 10.8	6 C 12.0	7 N 14.0	8 O 16.0	9 F 19.0	10 Ne 20.2
11 Na 23.0	12 Mg 24.3											13 Al 27.0	14 Si 28.1	15 P 31.0	16 S 32.1	17 Cl 35.5	18 Ar 39.9
19 K 39.1	20 Ca 40.1	21 Sc 45.0	22 Ti 47.9	23 V 50.9	24 Cr 52.0	25 Mn 54.9	26 Fe 55.8	27 Co 58.9	28 Ni 58.7	29 Cu 63.5	30 Zn 65.4	31 Ga 69.7	32 Ge 72.6	33 As 74.9	34 Se 79.0	35 Br 79.9	36 Kr 83.8
37 Rb 85.5	38 Sr 87.6	39 Y 88.9	40 Zr 91.2	41 Nb 92.9	42 Mo 95.9	43 Tc (98)	44 Ru 101.1	45 Rh 102.9	46 Pd 106.4	47 Ag 107.9	48 Cd 112.4	49 In 114.8	50 Sn 118.7	51 Sb 121.8	52 Te 127.6	53 I 126.9	54 Xe 131.3
55 Cs 132.9	56 Ba 137.3	57 *La 138.9	72 Hf 178.5	73 Ta 180.9	74 W 183.9	75 Re 186.2	76 Os 190.2	77 Ir 192.2	78 Pt 195.1	79 Au 197.0	80 Hg 200.6	81 Tl 204.4	82 Pb 207.2	83 Bi 209.0	84 Po (209)	85 At (210)	86 Rn (222)
87 Fr (223)	88 Ra 226.0	89 †Ac 227.0															

*Lanthanum Series

58 Ce 140.1	59 Pr 140.9	60 Nd 144.2	61 Pm (145)	62 Sm 150.4	63 Eu 152.0	64 Gd 157.3	65 Tb 158.9	66 Dy 162.5	67 Ho 164.9	68 Er 167.3	69 Tm 168.9	70 Yb 173.0	71 Lu 175.0

†Actinium Series

90 Th 232.0	91 Pa 231.0	92 U 238.0	93 Np 237.0	94 Pu (244)	95 Am (243)	96 Cm (247)	97 Bk (247)	98 Cf (251)	99 Es (252)	100 Fm (258)	101 Md (258)	102 No (259)	103 Lr (260)

DO NOT DETACH FROM BOOK.

GO ON TO THE NEXT PAGE.

Note: For all questions involving solutions and/or chemical equations, assume that the system is in pure water and at room temperature unless otherwise stated.

Part A

Directions: Each set of lettered choices below refers to the numbered questions or statements immediately following it. Select the one lettered choice that best answers each question or best fits each statement and fill in the corresponding oval on the answer sheet. A choice may be used once, more than once, or not at all in each set.

Questions 1–3 are based on the following energy diagrams.

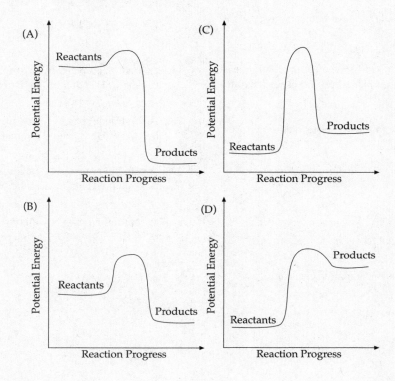

1. This reaction has the largest activation energy.

2. This is the most exothermic reaction.

3. This reaction has the largest positive value for ΔH.

GO ON TO THE NEXT PAGE.

Questions 4–6 refer to the following configurations.

(A) $1s^2\,2s^22p^6$
(B) $1s^2\,2s^22p^6\,3s^2$
(C) $1s^2\,2s^22p^6\,3s^23p^4$
(D) $1s^2\,2s^22p^6\,3s^23p^6$

4. The ground state configuration of an atom of a paramagnetic element.

5. The ground state configuration for both a potassium ion and a chloride ion.

6. An atom that has this ground-state electron configuration will have the smallest atomic radius of those listed above.

Questions 7–10 refer to the following molecules.

(A) H_2O
(B) SO_2
(C) NO_2
(D) O_2

7. In this molecule, oxygen forms sp^3 hybrid orbitals.

8. This molecule contains one unpaired electron.

9. This molecule contains no pi (π) bonds.

10. Oxygen has an oxidation state of zero in the molecule.

Questions 11–14 refer to the following solutions.

(A) A solution with a pH of 1
(B) A solution with a pH of greater than 1 and less than 7
(C) A solution with a pH of 7
(D) A solution with a pH of greater than 7 and less than 13

For CH_3COOH, $K_a = 1.8 \times 10^{-5}$

For NH_3, $K_b = 1.8 \times 10^{-5}$

11. A solution prepared by mixing equal volumes of 0.2-molar HCl and 0.2-molar NH_3.

12. A solution prepared by mixing equal volumes of 0.2-molar HNO_3 and 0.2-molar NaOH.

13. A solution prepared by mixing equal volumes of 0.2-molar HCl and 0.2-molar NaCl.

14. A solution prepared by mixing equal volumes of 0.2-molar CH_3COOH and 0.2-molar NaOH.

GO ON TO THE NEXT PAGE.

Part B

<u>Directions:</u> Each of the questions or incomplete statements below is followed by four suggested answers or completions. Select the one that is best in each case and fill in the corresponding oval on the answer sheet.

15. A pure sample of $KClO_3$ is found to contain 71 grams of chlorine atoms. What is the mass of the sample?

   (A) 122 grams
   (B) 170 grams
   (C) 209 grams
   (D) 245 grams

16. A chemist analyzed the carbon–carbon bond in C_2H_6 and found that it had a bond energy of 350 kJ/mol and a bond length of 1.5 angstroms. If the chemist performed the same analysis on the carbon–carbon bond in C_2H_2, how would the results compare?

   (A) The bond energy for C_2H_2 would be smaller, and the bond length would be shorter.
   (B) The bond energy for C_2H_2 would be greater, and the bond length would be longer.
   (C) The bond energy for C_2H_2 would be smaller, and the bond length would be longer.
   (D) The bond energy for C_2H_2 would be greater, and the bond length would be shorter.

17. Which of the following can function as both a Brønsted-Lowry acid and Brønsted-Lowry base?

   (A) H_2SO_4
   (B) HSO_3^-
   (C) SO_4^{2-}
   (D) H^+

18. Which of the following substances experiences the strongest attractive intermolecular forces?

   (A) H_2
   (B) N_2
   (C) CO_2
   (D) NH_3

19. A mixture of gases at equilibrium over water at 43°C contains 9.0 moles of nitrogen, 2.0 moles of oxygen, and 1.0 mole of water vapor. If the total pressure exerted by the gases is 780 mmHg, what is the vapor pressure of water at 43°C?

   (A)   65 mmHg
   (B)   130 mmHg
   (C)   260 mmHg
   (D)   580 mmHg

20. $$...MnO_4^- + ...e^- + ...H^+ \rightarrow ...Mn^{2+} + ...H_2O$$

   When the half-reaction above is balanced, what is the coefficient for H^+ if all the coefficients are reduced to the lowest whole number?

   (A) 3
   (B) 4
   (C) 5
   (D) 8

21. The boiling point of water is known to be lower at high elevations. This is because

   (A) hydrogen bonds are weaker at high elevations.
   (B) the heat of fusion is lower at high elevations.
   (C) the vapor pressure of water is higher at high elevations.
   (D) the atmospheric pressure is lower at high elevations.

22. A student examined 2.0 moles of an unknown carbon compound and found that the sample contained 48 grams of carbon, 64 grams of oxygen, and 8 grams of hydrogen. Which of the following could be the molecular formula of the compound?

   (A) CH_2O
   (B) CH_2OH
   (C) CH_3COOH
   (D) CH_3CO

GO ON TO THE NEXT PAGE.

23.
$$S(s) + O_2(g) \rightarrow SO_2(g) \qquad \Delta H = x$$
$$S(s) + \frac{3}{2} O_2(g) \rightarrow SO_3(g) \qquad \Delta H = y$$

Based on the information above, what is the standard enthalpy change for the following reaction?

$$2\ SO_2(g) + O_2(g) \rightarrow 2\ SO_3(g)$$

(A) $x - y$
(B) $y - x$
(C) $2x - 2y$
(D) $2y - 2x$

24. How much water must be added to a 50.0 ml solution of 0.60 M HNO_3 to produce a 0.40 M solution of HNO_3?

(A) 25 ml
(B) 33 ml
(C) 50 ml
(D) 75 ml

25. In which of the following equilibria would the concentrations of the products be increased if the volume of the system were decreased at constant temperature?

(A) $H_2(g) + Cl_2(g) \rightleftharpoons 2\ HCl(g)$
(B) $2\ CO(g) + O_2(g) \rightleftharpoons 2\ CO_3(g)$
(C) $NO(g) + O_3(g) \rightleftharpoons NO_2(g) + O_2(g)$
(D) $2\ HI(g) \rightleftharpoons H_2(g) + I_2(g)$

26. A 100 ml sample of 0.10 molar NaOH solution was added to 100 ml of 0.10 molar $H_3C_6H_5O_7$. After equilibrium was established, which of the ions listed below was present in the greatest concentration?

(A) $H_2C_6H_5O_7^-$
(B) $HC_6H_5O_7^{2-}$
(C) $C_6H_5O_7^{3-}$
(D) OH^-

27. The value of the equilibrium constant K_{eq} is greater than 1 for a certain reaction under standard state conditions. Which of the following statements must be true regarding the reaction?

(A) $\Delta G°$ is negative.
(B) $\Delta G°$ is positive.
(C) $\Delta G°$ is equal to zero.
(D) $\Delta G°$ is negative if the reaction is exothermic and positive if the reaction is endothermic.

28. The molecular formula for hydrated iron (III) oxide, or rust, is generally written as $Fe_2O_3 \cdot x\ H_2O$ because the water content in rust can vary. If a 1-molar sample of hydrated iron (III) oxide is found to contain 108 g of H_2O, what is the molecular formula for the sample?

(A) $Fe_2O_3 \cdot H_2O$
(B) $Fe_2O_3 \cdot 3\ H_2O$
(C) $Fe_2O_3 \cdot 6\ H_2O$
(D) $Fe_2O_3 \cdot 10\ H_2O$

29. In which of the following reactions does the greatest increase in entropy take place?

(A) $H_2O(l) \rightarrow H_2O(g)$
(B) $2\ NO(g) + O_2(g) \rightarrow 2\ NO_2(g)$
(C) $CaH_2(s) + H_2O(l) \rightarrow Ca(OH)_2(s) + H_2O(g)$
(D) $NH_4Cl(s) \rightarrow NH_3(g) + HCl(g)$

30. A boiling-water bath is sometimes used instead of a flame in heating objects. Which of the following could be an advantage of a boiling-water bath over a flame?

(A) The relatively high density of water will cause the object to become hot more quickly.
(B) The volume of boiling water remains constant over time.
(C) The temperature of boiling water remains constant at 100°C.
(D) The vapor pressure of boiling water is equal to zero.

31. The addition of a catalyst to a chemical reaction will bring about a change in which of the following characteristics of the reaction?

   I. The activation energy
  II. The enthalpy change
 III. The value of the equilibrium constant

(A) I only
(B) II only
(C) I and II only
(D) I and III only

GO ON TO THE NEXT PAGE.

32.
$$2 \, NO(g) + O_2(g) \rightarrow 2 \, NO_2(g)$$

The reaction above occurs by the following two-step process:

Step I: $NO(g) + O_2(g) \rightarrow NO_3(g)$

Step II: $NO_3(g) + NO(g) \rightarrow 2 \, NO_2(g)$

Which of the following is true of Step II if it is the rate-limiting step?

(A) Step II has a lower activation energy and occurs more slowly than Step I.
(B) Step II has a higher activation energy and occurs more slowly than Step I.
(C) Step II has a lower activation energy and occurs more quickly than Step I.
(D) Step II has a higher activation energy and occurs more quickly than Step I.

33.
$$C_3H_7OH + ...O_2 \rightarrow ...CO_2 + ...H_2O$$

One mole of C_3H_7OH underwent combustion as shown in the reaction above. How many moles of oxygen were required for the reaction?

(A) 2 moles

(B) 3 moles

(C) $\dfrac{7}{2}$ moles

(D) $\dfrac{9}{2}$ moles

34. Which of the following statements regarding atomic theory is NOT true?

(A) The Bohr model of the atom was based on Planck's quantum theory.
(B) Rutherford's experiments with alpha particle scattering led to the conclusion that positive charge was concentrated in an atom's nucleus.
(C) Heisenberg's uncertainty principle describes the equivalence of mass and energy.
(D) Millikan's oil drop experiment led to the calculation of the charge on an electron.

35.

Time (min)	[A] M
0	0.50
10	0.36
20	0.25
30	0.18
40	0.13

Reactant A underwent a decomposition reaction. The concentration of A was measured periodically and recorded in the chart above. Based on the data in the chart, which of the following statements is NOT true?

(A) The reaction is first order in [A].
(B) The reaction is first order overall.
(C) The rate of the reaction is constant over time.
(D) The half-life of reactant A is 20 minutes.

36.
$$N_2(g) + 3 \, H_2(g) \rightleftharpoons 2 \, NH_3(g) + energy$$

Which of the following changes to the equilibrium situation shown above will bring about an increase in the number of moles of NH_3 present at equilibrium?

I. Adding N_2 gas to the reaction chamber
II. Increasing the volume of the reaction chamber at constant temperature
III. Increasing the temperature of the reaction chamber at constant volume

(A) I only
(B) II only
(C) I and II only
(D) I and III only

37.
$$CH_4 + 2 \, O_2 \rightleftharpoons CO_2 + 2 \, H_2O$$

If 16 grams of CH_4 reacts with 16 grams of O_2 in the reaction shown above, which of the following will be true?

(A) The mass of H_2O formed will be twice the mass of CO_2 formed.
(B) Equal masses of H_2O and CO_2 will be formed.
(C) Equal numbers of moles of H_2O and CO_2 will be formed.
(D) The limiting reagent will be O_2.

GO ON TO THE NEXT PAGE.

38. Which of the following equilibrium expressions represents the hydrolysis of the CN⁻ ion?

(A) $K = \dfrac{[HCN][OH^-]}{[CN^-]}$

(B) $K = \dfrac{[CN^-][OH^-]}{[HCN]}$

(C) $K = \dfrac{[CN^-][H_3O^+]}{[HCN]}$

(D) $K = \dfrac{[HCN][H_3O^+]}{[CN^-]}$

39. Which of the following is true under any conditions for a reaction that is spontaneous at any temperature?

(A) ΔG, ΔS, and ΔH are all positive.
(B) ΔG, ΔS, and ΔH are all negative.
(C) ΔG and ΔS are positive, and ΔH is negative.
(D) ΔG and ΔH are negative, and ΔS is positive.

40. Which of the following pairs of compounds are isomers?

(A) HCOOH and CH_3COOH
(B) CH_3CH_2CHO and C_3H_7OH
(C) C_2H_5OH and CH_3OCH_3
(D) C_2H_4 and C_2H_6

41. A sample of an ideal gas confined in a rigid 5.00 liter container has a pressure of 363 mmHg at a temperature of 25°C. Which of the following expressions will be equal to the pressure of the gas if the temperature of the container is increased to 35°C?

(A) $\dfrac{(363)(35)}{(25)}$ mmHg

(B) $\dfrac{(363)(25)}{(35)}$ mmHg

(C) $\dfrac{(363)(308)}{(298)}$ mmHg

(D) $\dfrac{(363)(298)}{(308)}$ mmHg

42. If the temperature at which a reaction takes place is increased, the rate of the reaction will

(A) increase if the reaction is endothermic and decrease if the reaction is exothermic.
(B) decrease if the reaction is endothermic and increase if the reaction is exothermic.
(C) increase if the reaction is endothermic and increase if the reaction is exothermic.
(D) decrease if the reaction is endothermic and decrease if the reaction is exothermic.

43. The acid dissociation constant for HClO is 3.0×10^{-8}. What is the hydrogen ion concentration in a 0.12 M solution of HClO?

(A) $3.6 \times 10^{-8}\ M$
(B) $6.0 \times 10^{-8}\ M$
(C) $2.0 \times 10^{-5}\ M$
(D) $6.0 \times 10^{-5}\ M$

44. $2\,NO(g) + 2\,H_2(g)\ N_2(g) + 2\,H_2O(g)$

Which of the following is true regarding the relative molar rates of disappearance of the reactants and appearance of the products?

I. N_2 appears at the same rate that H_2 disappears.
II. H_2O appears at the same rate that NO disappears.
III. NO disappears at the same rate that H_2 disappears.

(A) I only
(B) I and II only
(C) II and III only
(D) I, II, and III

45. $2\,ZnS(s) + 3\,O_2(g) \rightarrow 2\,ZnO(s) + 2\,SO_2(g)$

If the reaction above took place at standard temperature and pressure, what was the volume of $O_2(g)$ required to produce 40.0 grams of ZnO(s)?

(A) $\dfrac{(40.0)(2)}{(81.4)(3)(22.4)}$ L

(B) $\dfrac{(40.0)(3)}{(81.4)(2)(22.4)}$ L

(C) $\dfrac{(40.0)(2)(22.4)}{(81.4)(3)}$ L

(D) $\dfrac{(40.0)(3)(22.4)}{(81.4)(2)}$ L

GO ON TO THE NEXT PAGE.

46. Which of the following salts will produce a colorless solution when added to water?

    (A) $Cu(NO_3)_2$
    (B) $NiCl_2$
    (C) $KMnO_4$
    (D) $ZnSO_4$

47. A beaker contains 300.0 ml of a 0.20 M $Pb(NO_3)_2$ solution. If 200.0 ml of a 0.20 M solution of $MgCl_2$ is added to the beaker, what will be the final concentration of Pb^{2+} ions in the resulting solution?

    (A) 0.020 M
    (B) 0.040 M
    (C) 0.080 M
    (D) 0.120 M

48. $$H_2(g) + I_2(g) \rightarrow 2\,HI(g)$$

    At 450°C the equilibrium constant, K_c, for the reaction shown above has a value of 50. Which of the following sets of initial conditions at 450°C will cause the reaction above to produce more H_2?

    I. $[HI]$ = 5-molar, $[H_2]$ = 1-molar,
         $[I_2]$ = 1-molar
    II. $[HI]$ = 10-molar, $[H_2]$ = 1-molar,
         $[I_2]$ = 1-molar
    III. $[HI]$ = 10-molar, $[H_2]$ = 2-molar,
         $[I_2]$ = 2-molar

    (A) I only
    (B) II only
    (C) I and II only
    (D) II and III only

49. Which of the following statements is true regarding sodium and potassium?

    (A) Sodium has a larger first ionization energy and a larger atomic radius.
    (B) Sodium has a larger first ionization energy and a smaller atomic radius.
    (C) Sodium has a smaller first ionization energy and a larger atomic radius.
    (D) Sodium has a smaller first ionization energy and a smaller atomic radius.

50. $$HCl(aq) + AgNO_3(aq) \rightarrow AgCl(s) + HNO_3(aq)$$

    One-half liter of a 0.20-molar HCl solution is mixed with one-half liter of a 0.40-molar solution of $AgNO_3$. A reaction occurs forming a precipitate as shown above. If the reaction goes to completion, what is the mass of AgCl produced?

    (A) 14 grams
    (B) 28 grams
    (C) 42 grams
    (D) 70 grams

51. $$H_2(g) + Cl_2(g) \rightarrow 2\,HCl(g)$$

    Based on the information given in the table below, what is H for the above reaction?

Bond	Average Bond Energy (kJ/mol)
H–H	440
Cl–Cl	240
H–Cl	430

    (A) −860 kJ
    (B) −620 kJ
    (C) −440 kJ
    (D) −180 kJ

52. The first ionization energy for magnesium is 730 kJ/mol. The third ionization energy for magnesium is 7700 kJ/mol. What is the most likely value for magnesium's second ionization energy?

    (A)   490 kJ/mol
    (B) 1,400 kJ/mol
    (C) 4,200 kJ/mol
    (D) 8,400 kJ/mol

53. Molten NaCl is electrolyzed with a constant current of 1.00 ampere. What is the shortest amount of time, in seconds, that it would take to produce 1.00 mole of solid sodium?
    (1 faraday = 96,500 coulombs)

    (A) 32,200 seconds
    (B) 48,300 seconds
    (C) 64,300 seconds
    (D) 96,500 seconds

GO ON TO THE NEXT PAGE.

54. How many moles of KCl must be added to 200 milliliters of a 0.5-molar NaCl solution to create a solution in which the concentration of Cl^- ion is 1.0-molar? (Assume the volume of the solution remains constant.)

    (A) 0.1 moles
    (B) 0.2 moles
    (C) 0.3 moles
    (D) 0.4 moles

55. A student placed solid barium oxalate in a beaker filled with distilled water and allowed it to come to equilibrium with its dissolved ions. The student then added a nickel (II) nitrate solution to the beaker to create the following equilibrium situation.

$$BaC_2O_4 + Ni^{2+} \rightleftharpoons Ba^{2+} + NiC_2O_4$$

    If the solubility product for BaC_2O_4 is 2×10^{-7} and the solubility product for NiC_2O_4 is 4×10^{-10}, which of the following gives the value of the equilibrium constant for the reaction above?

    (A) 2×10^3
    (B) 5×10^2
    (C) 5×10^{-2}
    (D) 2×10^{-3}

56. A 100-gram sample of pure $^{37}_{18}Ar$ decays by electron capture with a half-life of 35 days. How long will it take for 90 grams of $^{37}_{17}Cl$ to accumulate?

    (A) 31 days
    (B) 39 days
    (C) 78 days
    (D) 116 days

57. The solubility product, K_{sp}, of CaF_2 is 4×10^{-11}. Which of the following expressions is equal to the solubility of CaF_2?

    (A) $\sqrt{2 \times 10^{-11}}\ M$
    (B) $\sqrt[3]{4 \times 10^{-11}}\ M$
    (C) $\sqrt[3]{2 \times 10^{-11}}\ M$
    (D) $\sqrt[3]{1 \times 10^{-11}}\ M$

58. When excess hydroxide ions were added to 1.0 liter of $CaCl_2$ solution, $Ca(OH)_2$ precipitate was formed. If all of the calcium ions in the solution were precipitated in 7.4 grams of $Ca(OH)_2$, what was the initial concentration of the $CaCl_2$ solution?

    (A) 0.05-molar
    (B) 0.10-molar
    (C) 0.15-molar
    (D) 0.20-molar

59. When a solution of $KMnO_4$ was mixed with a solution of HCl, Cl_2 gas bubbles formed and Mn^{2+} ions appeared in the solution. Which of the following has occurred?

    (A) K^+ has been oxidized by Cl^-.
    (B) K^+ has been oxidized by H^+.
    (C) Cl^- has been oxidized by K^+.
    (D) Cl^- has been oxidized by MnO_4^-.

60. $2\ Cu^+(aq) + M(s) \rightarrow 2\ Cu(s) + M^{2+}(aq)$
    $$E^o = +0.92\ V$$

    $Cu^+(aq) + e^- \rightarrow Cu(s)$
    $$E^o = +0.52\ V$$

    Based on the reduction potentials given above, what is the standard reduction potential for the following half-reaction?

    $$M^{2+}(aq) + 2\ e^- \rightarrow M(s)$$

    (A) +0.40 V
    (B) +0.12 V
    (C) −0.12 V
    (D) −0.40 V

END OF SECTION I

CHEMISTRY
SECTION II
Time—1 hour and 30 minutes

General Instructions

Calculators, including those with programming and graphing capabilities, may be used in Part A. However, calculators with typewriter-style (QWERTY) keyboards are NOT permitted.

Pages containing a periodic table, the electrochemical series, and equations commonly used in chemistry will be available for your use.

You may write your answers with either a pen or a pencil. Be sure to write CLEARLY and LEGIBLY. If you make an error, you may save time by crossing it out rather than trying to erase it.

Write all your answers in the essay booklet. Number your answers as the questions are numbered in the examination booklet.

GO ON TO THE NEXT PAGE.

MATERIAL IN THE FOLLOWING TABLE AND IN THE TABLES ON THE NEXT 3 PAGES MAY BE USEFUL IN ANSWERING THE QUESTIONS IN THIS SECTION OF THE EXAMINATION.

PERIODIC CHART OF THE ELEMENTS

1 H 1.0																		2 He 4.0
3 Li 6.9	4 Be 9.0											5 B 10.8	6 C 12.0	7 N 14.0	8 O 16.0	9 F 19.0	10 Ne 20.2
11 Na 23.0	12 Mg 24.3											13 Al 27.0	14 Si 28.1	15 P 31.0	16 S 32.1	17 Cl 35.5	18 Ar 39.9
19 K 39.1	20 Ca 40.1	21 Sc 45.0	22 Ti 47.9	23 V 50.9	24 Cr 52.0	25 Mn 54.9	26 Fe 55.8	27 Co 58.9	28 Ni 58.7	29 Cu 63.5	30 Zn 65.4	31 Ga 69.7	32 Ge 72.6	33 As 74.9	34 Se 79.0	35 Br 79.9	36 Kr 83.8
37 Rb 85.5	38 Sr 87.6	39 Y 88.9	40 Zr 91.2	41 Nb 92.9	42 Mo 95.9	43 Tc (98)	44 Ru 101.1	45 Rh 102.9	46 Pd 106.4	47 Ag 107.9	48 Cd 112.4	49 In 114.8	50 Sn 118.7	51 Sb 121.8	52 Te 127.6	53 I 126.9	54 Xe 131.3
55 Cs 132.9	56 Ba 137.3	57 *La 138.9	72 Hf 178.5	73 Ta 180.9	74 W 183.9	75 Re 186.2	76 Os 190.2	77 Ir 192.2	78 Pt 195.1	79 Au 197.0	80 Hg 200.6	81 Tl 204.4	82 Pb 207.2	83 Bi 209.0	84 Po (209)	85 At (210)	86 Rn (222)
87 Fr (223)	88 Ra 226.0	89 †Ac 227.0															

*Lanthanum Series

58 Ce 140.1	59 Pr 140.9	60 Nd 144.2	61 Pm (145)	62 Sm 150.4	63 Eu 152.0	64 Gd 157.3	65 Tb 158.9	66 Dy 162.5	67 Ho 164.9	68 Er 167.3	69 Tm 168.9	70 Yb 173.0	71 Lu 175.0

†Actinium Series

90 Th 232.0	91 Pa 231.0	92 U 238.0	93 Np 237.0	94 Pu (244)	95 Am (243)	96 Cm (247)	97 Bk (247)	98 Cf (251)	99 Es (252)	100 Fm (258)	101 Md (258)	102 No (259)	103 Lr (260)

DO NOT DETACH FROM BOOK.

GO ON TO THE NEXT PAGE.

ADVANCED PLACEMENT CHEMISTRY EQUATIONS AND CONSTANTS

ATOMIC STRUCTURE

$$E = h\nu \qquad\qquad c = \lambda\nu$$

$$\lambda = \frac{h}{mv} \qquad\qquad p = mv$$

$$E_n = \frac{-2.178 \times 10^{-18}}{n^2} \text{ joule}$$

EQUILIBRIUM

$$K_a = \frac{[H^-][A^-]}{[HA]}$$

$$K_b = \frac{[OH^-][HB^+]}{[B]}$$

$$K_w = [OH^-][H^+] = 10^{-14} \text{ @ } 25°C$$

$$= K_a \times K_b$$

$$pH = -\log[H^+], \ pOH = -\log[OH^-]$$

$$14 = pH + pOH$$

$$pH = pK_a + \log\frac{[A^-]}{[HA]}$$

$$pOH = pK_b + \log\frac{[HB^+]}{[B]}$$

$$pK_a = -\log K_a, \ pK_b = -\log K_b$$

$$K_p = K_c(RT)^{\Delta n},$$

where Δn = moles product gas – moles reactant gas

THERMOCHEMISTRY / KINETICS

$$\Delta S° = \Sigma S°_{products} - \Sigma S°_{reactants}$$

$$\Delta H° = \Sigma \Delta H°_{f \ products} - \Sigma \Delta H°_{f \ reactants}$$

$$\Delta G° = \Sigma \Delta G°_{f \ products} - \Sigma \Delta G°_{f \ reactants}$$

$$\Delta G° = \Delta H° - T\Delta S°$$
$$= -RT \ln K = -2.303 \ RT \log K$$
$$= -nFE°$$

$$\Delta G = \Delta G° + RT \ln Q = \Delta G° + 2.303 \ RT \log Q$$
$$q = mc\Delta T$$
$$C_p = \frac{\Delta H}{\Delta T}$$

$$\ln[A]_t - \ln[A]_o = -kt$$

$$\frac{1}{[A]_t} - \frac{1}{[A]_o} = kt$$

$$\ln k = \frac{-E_a}{R}\left(\frac{1}{T}\right) + \ln A$$

E = energy $\qquad\qquad v$ = velocity

ν = frequency $\qquad\quad n$ = principal quantum number

λ = wavelength $\qquad\quad m$ = mass

p = momentum

Speed of light, $c = 3.00 \times 10^8 \text{ m s}^{-1}$

Planck's constant, $h = 6.63 \times 10^{-34} \text{ J s}$

Boltzmann's constant, $k = 1.38 \times 10^{-23} \text{ J K}^{-1}$

Avogadro's number $= 6.022 \times 10^{23} \text{ mol}^{-1}$

Electron charge, $e = -1.602 \times 10^{-19} \text{ coulomb}$

1 electron volt/atom $= 96.5 \text{ kJ mol}^{-1}$

Equilibrium Constants

K_a (weak acid)
K_b (weak base)
K_w (water)
K_p (gas pressure)
K_c (molar concentrations)

$S°$ = standard entropy
$H°$ = standard enthalpy
$G°$ = standard free energy
$E°$ = standard reduction potential
T = temperature
n = moles
m = mass
q = heat
c = specific heat capacity
C_p = molar heat capacity at constant pressure
E_a = activation energy
k = rate constant
A = frequency factor

Faraday's Constant, F = 96,500 coulombs per mole of electrons

Gas Constant, R = 8.31 J mol^{-1}K^{-1}
$\qquad\qquad$ = 0.0821 L atm mol^{-1}K^{-1}
$\qquad\qquad$ = 8.31 volt coulomb mol^{-1}K^{-1}

GO ON TO THE NEXT PAGE.

ADVANCED PLACEMENT CHEMISTRY EQUATIONS AND CONSTANTS

GASES, LIQUIDS, AND SOLUTIONS

$$PV = nRT$$

$$\left(P + \frac{n^2 a}{V^2}\right)(V - nb) = nRT$$

$$P_A = P_{total} \cdot X_A, \text{ where } X_A = \frac{\text{moles A}}{\text{total moles}}$$

$$P_{total} = P_A + P_B + P_C + \ldots$$

$$n = \frac{m}{M}$$

$$K = {}^\circ C + 273$$

$$\frac{P_1 V_1}{T_1} = \frac{P_2 V_2}{T_2}$$

$$D = \frac{m}{V}$$

$$U_{rms} = \sqrt{\frac{3kT}{m}} = \sqrt{\frac{3RT}{M}}$$

$$KE \text{ per molecule} = \frac{1}{2} mv^2$$

$$KE \text{ per mole} = \frac{3}{2} RT$$

$$\frac{r_1}{r_2} = \sqrt{\frac{M_2}{M_1}}$$

molarity, M = moles solute per liter solution
molality = moles solute per kilogram solvent

$$\Delta T_f = iK_f \times \text{molality}$$
$$\Delta T_b = iK_b \times \text{molality}$$
$$\pi = MRT$$
$$A = abc$$

OXIDATION REDUCTION; ELECTROCHEMISTRY

$$Q = \frac{[C]^c [D]^d}{[A]^a [B]^b} \text{ where } a\,A + b\,B \rightarrow c\,C + d\,D$$

$$I = \frac{q}{t}$$

$$E_{cell} = E^\circ_{cell} - \frac{RT}{nF} \ln Q = E^\circ_{cell} - \frac{0.0592}{n} \log Q \ @ \ 25^\circ C$$

$$\log K = \frac{nE^\circ}{0.0592}$$

P = pressure
V = volume
T = temperature
n = number of moles
D = density
m = mass
v = velocity

U_{rms} = root-mean-square speed
KE = kinetic energy
r = rate of effusion
M = molar mass
π = osmotic pressure
i = van't Hoff factor
K_f = molal freezing-point depression constant
K_b = molal boiling-point elevation constant
A = absorbance
a = molar absorptivity
b = path length
c = concentration
Q = reaction quotient
l = current (amperes)
q = charge (coulombs)
t = time (seconds)
E° = standard reduction potential
K = equilibrium constant

Gas constant, R = 8.31 J mol^{-1}K^{-1}
= 0.0821 L atm mol^{-1} K^{-1}
= 8.31 volt coulomb mol^{-1}K^{-1}
Boltzmann's constant, k = 1.38×10^{-23} J K^{-1}
K_f for H$_2$O = 1.86 K kg mol^{-1}
K_b for H$_2$O = 0.512 K kg mol^{-1}
1 atm = 760 mm Hg
= 760 torr
STP = 0.000°C and 1.000 atm
1 faraday, F = 96,500 coulumbs per mole
of electrons

GO ON TO THE NEXT PAGE.

STANDARD REDUCTION POTENTIALS IN AQUEOUS SOLUTION AT 25 C (in V)			
$F_2(g) + 2 e^-$	→	$2 F^-$	2.87
$Co^{3+} + e^-$	→	Co^{2+}	1.82
$Au^{3+} + 3e^-$	→	$Au(s)$	1.50
$Cl_2(g) + 2 e^-$	→	$2 Cl^-$	1.36
$O_2(g) + 4H^+ + 4 e^-$	→	$2 H_2O$	1.23
$Br_2(l) + 2 e^-$	→	$2 Br^-$	1.07
$2 Hg^{2+} + 2 e^-$	→	Hg_2^{2+}	0.92
$Hg^{2+} + 2 e^-$	→	$Hg(l)$	0.85
$Ag^+ + e^-$	→	$Ag(s)$	0.80
$Hg_2^{2+} + 2 e^-$	→	$2 Hg(l)$	0.79
$Fe^{3+} + e^-$	→	Fe^{2+}	0.77
$I_2(s) + 2 e^-$	→	$2 I^-$	0.53
$Cu^+ + e^-$	→	$Cu(s)$	0.52
$Cu^{2+} + 2 e^-$	→	$Cu(s)$	0.34
$Cu^{2+} + e^-$	→	Cu^+	0.15
$Sn^{4+} + 2 e^-$	→	Sn^{2+}	0.15
$S(s) + 2 H^+ + 2 e^-$	→	H_2S	0.14
$2 H^+ + 2 e^-$	→	$H_2(g)$	0.00
$Pb^{2+} + 2 e^-$	→	$Pb(s)$	−0.13
$Sn^{2+} + 2 e^-$	→	$Sn(s)$	−0.14
$Ni^{2+} + 2 e^-$	→	$Ni(s)$	−0.25
$Co^{2+} + 2 e^-$	→	$Co(s)$	−0.28
$Tl^+ + e^-$	→	$Tl(s)$	−0.34
$Cd^{2+} + 2 e^-$	→	$Cd(s)$	−0.40
$Cr^{3+} + e^-$	→	Cr^{2+}	−0.41
$Fe^2 + 2 e^-$	→	$Fe(s)$	−0.44
$Cr^{3+} + 3 e^-$	→	$Cr(s)$	−0.74
$Zn^{2+} + 2 e^-$	→	$Zn(s)$	−0.76
$Mn^{2+} + 2 e^-$	→	$Mn(s)$	−1.18
$Al^{3+} + 3 e^-$	→	$Al(s)$	−1.66
$Be^{2+} + 2 e^-$	→	$Be(s)$	−1.70
$Mg^{2+} + 2 e^-$	→	$Mg(s)$	−2.37
$Na^+ + e^-$	→	$Na(s)$	−2.71
$Ca^{2+} + 2 e^-$	→	$Ca(s)$	−2.87
$Sr^{2+} + 2 e^-$	→	$Sr(s)$	−2.89
$Ba^{2+} + 2 e^-$	→	$Ba(s)$	−2.90
$Rb^+ + e^-$	→	$Rb(s)$	−2.92
$K^+ + e^-$	→	$K(s)$	−2.92
$Cs^+ + e^-$	→	$Cs(s)$	−2.92
$Li^+ + e^-$	→	$Li(s)$	−3.05

GO ON TO THE NEXT PAGE.

CHEMISTRY

Section II

(Total time—90 minutes)

Part A

YOU MAY USE YOUR CALCULATOR FOR PART A

THE METHODS USED AND THE STEPS INVOLVED IN ARRIVING AT YOUR ANSWERS MUST BE SHOWN CLEARLY. It is to your advantage to do this since you may obtain partial credit if you do, and you will receive little or no credit if you do not. Attention should be paid to significant figures.

Be sure to write your answers in the space provided following each question.

Answer Questions 1, 2, and 3.

1. A 0.20-molar solution of acetic acid, $HC_2H_3O_2$, at a temperature of 25°C, has a pH of 2.73.

   (a) Calculate the hydroxide ion concentration, $[OH^-]$.

   (b) What is the value of the acid ionization constant, K_a, for acetic acid at 25°C?

   (c) How many moles of sodium acetate must be added to 500.0 ml of a 0.200-molar solution of acetic acid to create a buffer with a pH of 4.00? Assume that the volume of the solution is not changed by the addition of sodium acetate.

   (d) In a titration experiment, 100.0 ml of sodium hydroxide solution was added to 200 ml of a 0.400-molar solution of acetic acid until the equivalence point was reached. What was the pH at the equivalence point?

   (e) The pK_a values for several indicators are given in the table below. Which of the indicators on the table is most suitable for this titration? Justify your answer.

Indicator	pK_a
Thymol Blue	2
Bromcresol Purple	6
Phenolphthalein	9

GO ON TO THE NEXT PAGE.

2.

$$2\,NO(g) + Cl_2(g) \rightarrow 2\,NOCl(g)$$

The following data were collected for the reaction above. All of the measurements were taken at a temperature of 263 K.

Experiment	Initial [NO] (M)	Initial [Cl$_2$] (M)	Initial rate of disappearance of Cl$_2$ (M/min)
1	0.15	0.15	0.60
2	0.15	0.30	1.2
3	0.30	0.15	2.4
4	0.25	0.25	?

(a) Write the expression for the rate law for the reaction above.
(b) Calculate the value of the rate constant for the above reaction, and specify the units.
(c) What is the initial rate of appearance of NOCl in experiment 2?
(d) What is the initial rate of disappearance of Cl$_2$ in experiment 4?
(e) Each of the experimental trials took place in a closed container. Explain or calculate each of the following:
   (i) What was the partial pressure due to NO(g) at the start of experiment 1?
   (ii) What was the total pressure at the start of experiment 1? Assume that no NOCl is present.

3. A titration experiment was conducted to determine the pH of a known volume of a strong monoprotic acid solution. A hydroxide solution of known concentration was poured into a buret and then titrated into the acid solution. The volume of hydroxide solution titrated into the acid solution was measured at the equivalence point and used to calculate the concentration of the acid solution. Which of the following situations would cause an error in the calculated value of the pH of the acid solution? Explain each answer.

(a) The buret was rinsed with the hydroxide solution before the solution was poured into the buret.
(b) The experimenter did not notice that a few drops of hydroxide solution spattered outside the acid solution container during the titration.
(c) The buret was rinsed with distilled water before the hydroxide solution was added.
(d) The experimenter read the hydroxide solution level from the top of the fluid instead of the bottom of the meniscus both before the titration and at the equivalence point.
(e) Some hydroxide solution was spilled while the experimenter was pouring it into the buret.

GO ON TO THE NEXT PAGE.

Part B
CALCULATORS MAY BE USED FOR PART B

4.

$$CH_4(g) + 2\,O_2(g) \rightarrow CO_2(g) + 2\,H_2O(l)$$

The above reaction for the combustion of methane gas has a standard entropy change, $\Delta S°$, with a value of -242.7 J/mol-K. The following data are also available:

Compound	ΔH_f (kJ/mol)
$CH_4(g)$	-74.8
$H_2O\ (l)$	-285.9
$CO_2\ (g)$	-393.5

(a) What are the values of $\Delta H°_f$ and $\Delta G°_f$ for $O_2(g)$?

(b) Calculate the standard change in enthalpy, $\Delta H°$, for the combustion of methane.

(c) Calculate the standard free energy change, $\Delta G°$, for the combustion of methane.

(d) How would the value of $\Delta S°$ for the reaction be affected if the water produced in the combustion remained in the gas phase?

(e) A 20.0-gram sample of $CH_4(g)$ underwent combustion in a bomb calorimeter with excess oxygen gas.
   (i) Calculate the mass of carbon dioxide produced.
   (ii) Calculate the heat released by the reaction.

GO ON TO THE NEXT PAGE.

Answer Question 5 below.

5. You will be given three chemical reactions below. In part (i), write the balanced equation for the reaction, leaving coefficients in terms of lowest whole numbers. Then answer the question pertaining to that reaction in part (ii). For each of the following three reactions, assume that solutions are aqueous unless it says otherwise. Substances in solutions should be represented as ions if these substances are extensively ionized. Omit formulas for ions or molecules that are not affected by the reaction. Only equations inside the answer boxes will be graded.

EXAMPLE:

A piece of solid zinc is placed in a solution of silver(I) acetate

(i) Write the balanced equation in this box:

$$Zn + 2Ag^+ \rightarrow Zn^{2+} + 2Ag$$

(ii) Which substance is reduced in the reaction?

Ag^+ is reduced

(a) Sulfur dioxide gas is bubbled through cold water.

(i) Write the balanced equation in this box:

(ii) As the reaction progresses, will the hydroxide concentration in the solution increase, decrease, or remain the same?

GO ON TO THE NEXT PAGE.

(b) Ethane is burned in air.

(i) Write the balanced equation in this box:

(ii) If the enthalpy change for the reaction was measured to be –3,100 kJ and the total heat of formation for all products formed is –3,300 kJ, what is the approximate heat of formation of ethane measured in kJ/mole?

_____

_____

(c) Chlorine gas is bubbled through a solution of sodium bromide.

(i) Write the balanced equation in this box:

(ii) Which substance is reduced in this reaction?

_____

_____

GO ON TO THE NEXT PAGE.

Answer Question 6 and Question 7.

Answering these questions provides an opportunity to demonstrate your ability to present your material in logical, coherent, and convincing English. Your responses will be judged on the basis of accuracy and importance of the detail cited and on the appropriateness of the descriptive material used. Specific answers are preferable to broad, diffuse responses. Illustrative examples and equations may be helpful.

6. Oxygen is found in the atmosphere as a diatomic gas, O_2, and as ozone, O_3. Ozone has a dipole moment of 0.5 debye. Oxygen gas has a dipole moment of zero.

   (a)   Draw the Lewis dot structures for both molecules.

   (b)   Use the principles of bonding and molecular structure to account for the fact that ozone has a higher boiling point than diatomic oxygen.

   (c)   Use the principles of bonding and molecular structure to account for the fact that ozone is more soluble than diatomic oxygen in water.

   (d)   Explain why the two bonds in O_3 are of equal length and are longer than the bond length of the bond in diatomic oxygen.

   (e)   Elemental oxygen is strongly affected by a magnetic field. Explain why.

   (f)   For the equilibrium reaction below at 25°C and 1 atm, $K_{eq} = 10^{-57}$.

   $$3\, O_2(g) \leftrightarrow 2\, O_3(g)$$

   Which form of oxygen is more abundant under normal conditions? Explain.

GO ON TO THE NEXT PAGE.

7. $$NaC_2H_3O_2, \ Ba(NO_3)_2, \ KCl$$

(a) Aqueous solutions of equal concentration of the three compounds listed above are prepared. What would an experimenter expect to observe when each of the following procedures is performed on each of the solutions?

    (i) The pH of each solution is measured.

    (ii) SO_4^{2-} ions are introduced into each solution.

    (iii) The freezing point of each solution is measured, and the three temperatures are compared.

    (iv) Each solution is subjected to a flame test.

(b) At 25°C and 1.0 atmosphere pressure, a balloon contains a mixture of four ideal gases: oxygen, nitrogen, carbon dioxide, and helium. The partial pressure due to each gas is 0.25 atmosphere. Use the ideas of kinetic molecular theory to answer each of the following questions.

    (i) Rank the gases in increasing order of average molecular velocity.

    (ii) How would the volume of the balloon be affected if the temperature of the gases in the balloon were increased at constant pressure?

    (iii) What changes to the temperature and pressure of the gases would cause deviation from ideal behavior, and which gas would be most affected?

STOP

END OF EXAM

Chapter 11
Practice Test 1
Answers and
Explanations

ANSWER KEY EXAM 1

1. C		31. A	
2. A		32. B	
3. D		33. D	
4. C		34. C	
5. D		35. C	
6. A		36. A	
7. A		37. D	
8. C		38. A	
9. A		39. D	
10. D		40. C	
11. B		41. C	
12. C		42. C	
13. A		43. D	
14. D		44. C	
15. D		45. D	
16. D		46. D	
17. B		47. B	
18. D		48. B	
19. A		49. B	
20. D		50. A	
21. D		51. D	
22. C		52. B	
23. D		53. D	
24. A		54. A	
25. B		55. B	
26. A		56. D	
27. A		57. D	
28. C		58. B	
29. D		59. D	
30. C		60. D	

SECTION I—MULTIPLE CHOICE

Questions 1–3 are based on the following energy diagrams.

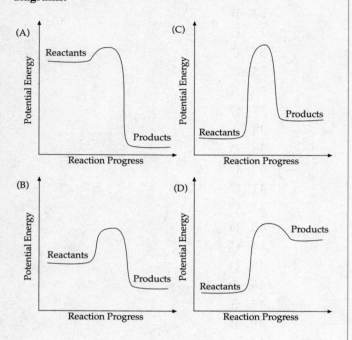

(A)   (C)

(B)   (D)

1. This reaction has the largest activation energy.

1.

(C)

C   This reaction has the largest rise from the energy
level of the reactants to the peak energy that must be
overcome in order for the reaction to proceed.

QUESTIONS	EXPLANATIONS

QUESTIONS

2. This is the most exothermic reaction.

3. This reaction has the largest positive value for ΔH.

EXPLANATIONS

2.

(A)

A   This reaction has the largest energy drop from the level of the reactants to the level of the products. That makes it the most exothermic.

3.

(D)

D   This reaction has the largest rise in energy from the level of the reactants to the level of the products. That makes it the most endothermic and gives it the largest positive value for ΔH.

QUESTIONS	EXPLANATIONS

Questions 4–6 refer to the following configurations.

(A) $1s^2\,2s^22p^6$
(B) $1s^2\,2s^22p^6\,3s^2$
(C) $1s^2\,2s^22p^6\,3s^23p^6$
(D) $1s^2\,2s^22p^6\,3s^23p^4$

4. The ground state configuration of an atom of a paramagnetic element.

4. **C** A paramagnetic element is one whose electrons are not completely spin-paired. For choice (C), sulfur, Hund's rule says that the first three electrons in the $3p$ subshell will each occupy an empty orbital. The fourth electron will pair up, leaving two electrons unpaired. All of the other choices have completed subshells, so all the electrons will be spin-paired.

5. The ground state configuration for both a potassium ion and a chloride ion.

5. **D** A potassium atom loses an electron when it ionizes; a chlorine atom gains an electron. Both end up with the same ground state electron configuration as an argon atom, given in choice (D).

6. An atom that has this ground-state electron configuration will have the smallest atomic radius of those listed above.

6. **A** This atom (neon) has electrons in only two shells. All of the other choices have electrons at higher energy levels, farther away from the nucleus.

QUESTIONS	EXPLANATIONS

Questions 7–10 refer to the following molecules.

(A) H_2O
(B) SO_2
(C) NO_2
(D) O_2

7. In this molecule, oxygen forms sp^3 hybrid orbitals.

7. **A** The Lewis dot structure for water is shown below.

The central oxygen atom forms sp^3 hybrid orbitals, resulting in a tetrahedral structure. The central oxygen atom has two unbonded electron pairs, which gives the molecule a bent shape.

8. This molecule contains one unpaired electron.

8. **C** NO_2 has an odd number (17) of valence electrons, so when we draw the Lewis dot structure, there must be an unpaired electron.

9. This molecule contains no pi (π) bonds.

9. **A** Water has only single bonds or sigma (σ) bonds. All of the other molecules have at least one double bond. The second bond in a double bond is a pi (π) bond.

10. Oxygen has an oxidation state of zero in this molecule.

10. **D** The oxidation state for an uncombined element is zero.

QUESTIONS	EXPLANATIONS

Questions 11–14 refer to the following solutions.

    (A)   A solution with a pH of 1
    (B)   A solution with a pH of greater than 1 and less than 7
    (C)   A solution with a pH of 7
    (D)   A solution with a pH of greater than 7 and less than 13

For CH_3COOH, $K_a = 1.8 \times 10^{-5}$
For NH_3, $K_b = 1.8 \times 10^{-5}$

11. A solution prepared by mixing equal volumes of 0.2-molar HCl and 0.2-molar NH_3.

    11. B This is a mixture of a strong acid and a weak base, so at the equivalence point, the mixture will be acidic with a pH less than 7.

12. A solution prepared by mixing equal volumes of 0.2-molar HNO_3 and 0.2-molar NaOH.

    12. C This is a mixture of a strong acid and a strong base, so at the equivalence point, they will have completely neutralized each other and only salt water will be left. Therefore, the solution will be neutral, and the pH will be 7.

13. A solution prepared by mixing equal volumes of 0.2-molar HCl and 0.2-molar NaCl.

    13. A Both HCl and NaCl dissociate completely, but NaCl will have no effect on the pH of the solution.

    Because we are doubling the volume of the HCl solution by adding the salt water, the concentration of HCl will be cut in half, to 0.1-molar. HCl is a strong acid, so $[H^+] = 0.1$-molar and pH $= -\log[H^+] = 1$.

14. A solution prepared by mixing equal volumes of 0.2-molar CH_3COOH and 0.2-molar NaOH.

    14. D This is a mixture of a weak acid and a strong base, so at the equivalence point, the mixture will be basic with a pH greater than 7.

QUESTIONS	EXPLANATIONS

15. A pure sample of $KClO_3$ is found to contain 71 grams of chlorine atoms. What is the mass of the sample?

   (A) 122 grams
   (B) 170 grams
   (C) 209 grams
   (D) 245 grams

15. **D** Remember: moles $= \dfrac{grams}{MW}$

So moles of chlorine $= \dfrac{(71 g)}{(35.5 \text{ g/mol})} = 2$ moles

If there are 2 moles of chlorine, there must be 2 moles of $KClO_3$.

The molecular weight of $KClO_3$ is 122.5.

So grams of $KClO_3$ = (moles)(MW)

   = (2 mol)(122.5 g/mol)
   ≐ 245 grams

16. A chemist analyzed the carbon–carbon bond in C_2H_6 and found that it had a bond energy of 350 kJ/mol and a bond length of 1.5 angstroms. If the chemist performed the same analysis on the carbon–carbon bond in C_2H_2, how would the results compare?

   (A) The bond energy for C_2H_2 would be smaller, and the bond length would be shorter.
   (B) The bond energy for C_2H_2 would be greater, and the bond length would be longer.
   (C) The bond energy for C_2H_2 would be smaller, and the bond length would be longer.
   (D) The bond energy for C_2H_2 would be greater, and the bond length would be shorter.

16. **D** First, you need to know that C_2H_6 is an alkane, which has a single C–C bond, and C_2H_2 is an alkyne, which has a triple C–C bond. Once you realize that, you know that a triple bond will be stronger and shorter than a single bond.

17. Which of the following can function as both a Brønsted-Lowry acid and Brønsted-Lowry base?

   (B) H_2SO_4
   (C) HSO_3^-
   (D) SO_4^{2-}
   (E) H^+

17. **B** HSO_3^- is amphoteric.

It can act as a Brønsted-Lowry acid, giving up a proton to become SO_3^{2-}.

It can act as a Brønsted-Lowry base, gaining a proton to become H_2SO_3.

H_2SO_4 can give up only protons.

SO_4^{2-} can gain only protons.

H^+ *is* a proton.

QUESTIONS	EXPLANATIONS

18. Which of the following substances experiences the strongest attractive intermolecular forces?

   (A) H_2
   (B) N_2
   (C) CO_2
   (D) NH_3

18. **D** NH_3 is the only molecule listed that undergoes hydrogen bonding. In fact, it is the only polar molecule listed.

19. A mixture of gases at equilibrium over water at 43°C contains 9.0 moles of nitrogen, 2.0 moles of oxygen, and 1.0 mole of water vapor. If the total pressure exerted by the gases is 780 mmHg, what is the vapor pressure of water at 43°C?

   (A)   65 mmHg
   (B) 130 mmHg
   (C) 260 mmHg
   (D) 580 mmHg

19. **A** From Dalton's law, partial pressure of a gas in a sample is directly proportional to its molar quantity, so if $\dfrac{1}{12}$ of the gas in the sample is water vapor, then $\dfrac{1}{12}$ of the total pressure will be due to water vapor. So the partial pressure of water vapor is (780) = 65 mmHg.
The gases are at equilibrium, so the partial pressure of the water vapor will be the same as the vapor pressure of the water.

20. $...MnO_4^- + ...e^- + ...H^+ \rightarrow ...Mn^{2+} + ...H_2O$

   When the half-reaction above is balanced, what is the coefficient for H^+ if all the coefficients are reduced to the lowest whole number?

   (A) 3
   (B) 4
   (C) 5
   (D) 8

20. **D** Backsolve. Start with (C). If there are 5 H^+, there can't be a whole number coefficient for H_2O, so (C) is wrong. You should also be able to see that the answer can't be an odd number, so (A) is also wrong.
Try (D).
If there are 8 H^+, then there are 4 H_2O.
If there are 4 H_2O, then there is 1 MnO_4^-.
If there is 1 MnO_4^-, then there is 1 Mn^{2+}.
Mn^{7+} (in MnO_4^-) is reduced to Mn^{2+}, so there are 5 e^-.
These are the lowest whole number coefficients, so (D) is correct.

21. The boiling point of water is known to be lower at high elevations. This is because

   (A) hydrogen bonds are weaker at high elevations.
   (B) the heat of fusion is lower at high elevations.
   (C) the vapor pressure of water is higher at high elevations.
   (D) the atmospheric pressure is lower at high elevations.

21. **D** Vapor pressure increases with increasing temperature. Water boils when its vapor pressure is equal to the atmospheric pressure. So if the atmospheric pressure is lowered, then water will boil at a lower temperature.

QUESTIONS	EXPLANATIONS

22. A student examined 2.0 moles of an unknown carbon compound and found that the sample contained 48 grams of carbon, 64 grams of oxygen, and 8 grams of hydrogen. Which of the following could be the molecular formula of the compound?

(A) CH_2O
(B) CH_2OH
(C) CH_3COOH
(D) CH_3CO

22. C You can find the number of moles of C, H, and O.

$$Moles = \frac{grams}{MW}$$

$$Moles\ of\ carbon = \frac{(48\,g)}{(12\,g\,/\,mol)} = 4\ moles$$

$$Moles\ of\ hydrogen = \frac{(8\,g)}{(1\,g\,/\,mol)} = 8\ moles$$

$$Moles\ of\ oxygen = \frac{(64\,g)}{(16\,g\,/\,mol)} = 4\ moles$$

The student had 2 moles of the compound, so 1 mole would contain half as much stuff. That's 2 moles of C, 4 moles of H, and 2 moles of O. That corresponds to CH_3COOH, acetic acid. By the way, choice (A), CH_2O, is wrong because it's the empirical formula, not the molecular formula.

23.

$$S(s) + O_2(g) \rightarrow SO_2(g) \qquad \Delta H = x$$

$$S(s) + \frac{3}{2}\,O_2(g) \rightarrow SO_3(g) \qquad \Delta H = y$$

Based on the information above, what is the standard enthalpy change for the following reaction?

$$2\,SO_2(g) + O_2(g) \rightarrow 2\,SO_3(g)$$

(A) $x - y$
(B) $y - x$
(C) $2x - 2y$
(D) $2y - 2x$

23. D The equations given on top give the heats of formation of all the reactants and products (remember: the heat of formation of O_2, an element in its most stable form, is zero).

$$\Delta H^\circ_{reaction} = \Delta H^\circ_{products} - \Delta H^\circ_{reactants}$$

First, the products.

From SO_3, we get $2y$. That's it for the products.

Now the reactants.

From SO_2, we get $2x$. The heat of formation of O_2 is defined to be zero, so that's it for the reactants.

ΔH for the reaction = $2y - 2x$

24. How much water must be added to a 50.0 ml solution of 0.60 M HNO_3 to produce a 0.40 M solution of HNO_3?

(A) 25 ml
(B) 33 ml
(C) 50 ml
(D) 75 ml

24. A Remember: moles = (molarity)(volume)

The number of moles of HNO_3 will remain constant during the dilution.

Moles of HNO_3 = (0.60 M)(50.0 ml)
= 30 millimoles

Now we can find how much water it will take to create a 0.40 M solution with 30 millimoles of HNO_3.

$$Volume = \frac{moles}{molarity} = \frac{30\ millimoles}{0.40\ M} = 75\ ml$$

But 75 ml isn't the answer. We started with 50 ml of solution and ended up with 75 ml of solution, so we must have added 25 ml of water.

QUESTIONS	EXPLANATIONS

25. In which of the following equilibria would the concentrations of the products be increased if the volume of the system were decreased at constant temperature?

 (A) $H_2(g) + Cl_2(g) \rightleftharpoons 2\,HCl(g)$
 (B) $2\,CO(g) + O_2(g) \rightleftharpoons 2\,CO_3(g)$
 (C) $NO(g) + O_3(g) \rightleftharpoons NO_2(g) + O_2(g)$
 (D) $2\,HI(g) \rightleftharpoons H_2(g) + I_2(g)$

25. **B** According to Le Châtelier's law, the equilibrium will shift to counteract any stress that is placed on it.

 If the volume is decreased, the equilibrium must shift toward the side with fewer moles of gas. Only choice (B) has fewer moles of gas on the product side (2 moles) than on the reactant side (3 moles).

26. A 100 ml sample of 0.10 molar NaOH solution was added to 100 ml of 0.10 molar $H_3C_6H_5O_7$. After equilibrium was established, which of the ions listed below was present in the greatest concentration?

 (A) $H_2C_6H_5O_7^-$
 (B) $HC_6H_5O_7^{2-}$
 (C) $C_6H_5O_7^{3-}$
 (D) OH^-

26. **A** Just enough NaOH has been added to neutralize all of the $H_3C_6H_5O_7$ in the reaction shown below.

 $$H_3C_6H_5O_7(aq) \rightarrow H^+(aq) + H_2C_6H_5O_7^-(aq)$$

 So the $H_3C_6H_5O_7(aq)$ is gone, and the $H_2C_6H_5O_7^-(aq)$ will be present in the greatest concentration because it won't dissociate much further.

27. The value of the equilibrium constant K_{eq} is greater than 1 for a certain reaction under standard state conditions. Which of the following statements must be true regarding the reaction?

 (A) ΔG° is negative.
 (B) ΔG° is positive.
 (C) ΔG° is equal to zero.
 (D) ΔG° is negative if the reaction is exothermic and positive if the reaction is endothermic.

27. **A** From the relationship $\Delta G^\circ = -RT \ln K$, we can see that if K is greater than 1, then $\ln K$ must be greater than 1, which means that ΔG° must be less than zero.

28. The molecular formula for hydrated iron (III) oxide, or rust, is generally written as $Fe_2O_3 \cdot x\,H_2O$ because the water content in rust can vary. If a 1-molar sample of hydrated iron (III) oxide is found to contain 108 g of H_2O, what is the molecular formula for the sample?

 (A) $Fe_2O_3 \cdot H_2O$
 (B) $Fe_2O_3 \cdot 3\,H_2O$
 (C) $Fe_2O_3 \cdot 6\,H_2O$
 (D) $Fe_2O_3 \cdot 10\,H_2O$

28. **C** $Moles = \dfrac{grams}{MW}$

 $Moles\ of\ H_2O = \dfrac{(108\ g)}{(18\ g/mol)} = 6\ moles$

 So if 1 mole of hydrate contains 6 moles of H_2O, then its formula must be $Fe_2O_3 \cdot 6\,H_2O$.

QUESTIONS	EXPLANATIONS
29. In which of the following reactions does the greatest increase in entropy take place? (A) $H_2O(l) \rightarrow H_2O(g)$ (B) $2\,NO(g) + O_2(g) \rightarrow 2\,NO_2(g)$ (C) $CaH_2(s) + H_2O(l) \rightarrow Ca(OH)_2(s) + H_2(g)$ (D) $NH_4Cl(s) \rightarrow NH_3(g) + HCl(g)$	29. **D** In choice (D), the number of moles goes from 1 to 2, and the phase changes from solid to gas, making this a pretty big entropy change. Entropy is also increasing in (A) and (C), but in both cases, the number of moles stays the same and the phase change is only from liquid to gas. The entropy decreases in (B).
30. A boiling-water bath is sometimes used instead of a flame in heating objects. Which of the following could be an advantage of a boiling-water bath over a flame? (A) The relatively high density of water will cause the object to become hot more quickly. (B) The volume of boiling water remains constant over time. (C) The temperature of boiling water remains constant at 100°C. (D) The vapor pressure of boiling water is equal to zero.	30. **C** The fact that boiling water maintains a constant temperature of 100°C is useful when a relatively low constant temperature is required. None of the other choices is a true statement.

31. The addition of a catalyst to a chemical reaction will bring about a change in which of the following characteristics of the reaction?
 	
  I. The activation energy
  II. The enthalpy change
  III. The value of the equilibrium constant
 	
(A) I only
(B) II only
(C) I and II only
(D) I and III only
	31. **A** The addition of a catalyst lowers the activation energy of a reaction, making it easier for the reaction to proceed, so (I) is correct.
	
	Adding a catalyst has no effect on the enthalpy change or equilibrium conditions of a reaction, so (II) and (III) are wrong.

32. $$2\,NO(g) + O_2(g) \rightarrow 2\,NO_2(g)$$ The reaction above occurs by the following two-step process: Step I: $NO(g) + O_2(g) \rightarrow NO_3(g)$ Step II: $NO_3(g) + NO(g) \rightarrow 2\,NO_2(g)$ Which of the following is true of Step II if it is the rate-limiting step? (A) Step II has a lower activation energy and occurs more slowly than Step I. (B) Step II has a higher activation energy and occurs more slowly than Step I. (C) Step II has a lower activation energy and occurs more quickly than Step I. (D) Step II has a higher activation energy and occurs more quickly than Step I.	32. **B** The rate-limiting step in a reaction is the slowest step in the process. When a reaction occurs slowly, it is because very few collisions among reactant molecules have enough energy to overcome the high activation energy.

QUESTIONS	EXPLANATIONS

33.

$$C_3H_7OH + ...O_2 \rightarrow ...CO_2 + ...H_2O$$

One mole of C_3H_7OH underwent combustion as shown in the reaction above. How many moles of oxygen were required for the reaction?

(A) 2 moles

(B) 3 moles

(C) $\dfrac{7}{2}$ moles

(D) $\dfrac{9}{2}$ moles

33. D Backsolve. Start at (C).

Instead of using 1 as the coefficient for C_3H_7OH and $\dfrac{7}{2}$ for O_2, use 2 for C_3H_7OH and 7 for O_2. This won't change your result, and it will make the math easier.

If there are 2 moles of C_3H_7OH, then there must be 6 moles of CO_2 and 8 moles of H_2O.

That gives us 16 O's in the reactants and 20 O's in the products. So (C) is wrong and we should pick a larger number to put more O's on the reactant side. Try (D).

There are still 2 moles of C_3H_7OH, so there must still be 6 moles of CO_2 and 8 moles of H_2O. Now there are 9 moles of O_2. Now we have 20 O's in the reactants and 20 O's in the products, so (D) is the correct answer.

34. Which of the following statements regarding atomic theory is not true?

(A) The Bohr model of the atom was based on Planck's quantum theory.
(B) Rutherford's experiments with alpha particle scattering led to the conclusion that positive charge was concentrated in an atom's nucleus.
(C) Heisenberg's uncertainty principle describes the equivalence of mass and energy.
(D) Millikan's oil drop experiment led to the calculation of the charge on an electron.

34. C The Heisenberg uncertainty principle states that both the momentum and location of an electron can never be known with absolute certainty. It was Einstein who described the equivalence of mass and energy.

35.

Time (min)	[A] M
0	0.50
10	0.36
20	0.25
30	0.18
40	0.13

Reactant A underwent a decomposition reaction. The concentration of A was measured periodically and recorded in the chart above. Based on the data in the chart, which of the following statements is NOT true?

(A) The reaction is first order in [A].
(B) The reaction is first order overall.
(C) The rate of the reaction is constant over time.
(D) The half-life of reactant A is 20 minutes.

35. C The reaction rate is NOT constant over time, so this is the statement that is not true. In fact, the reaction rate is gradually decreasing as the concentration of reactant A decreases. All of the other statements are true. You can see from the chart that the half-life of A is 20 minutes. The other three statements must be true based on the fact that [A] is decreasing exponentially.

QUESTIONS	EXPLANATIONS

36.

$$N_2(g) + 3\,H_2(g) \rightleftharpoons 2\,NH_3(g) + energy$$

Which of the following changes to the equilibrium situation shown above will bring about an increase in the number of moles of NH_3 present at equilibrium?

   I. Adding N_2 gas to the reaction chamber

   II. Increasing the volume of the reaction chamber at constant temperature

   III. Increasing the temperature of the reaction chamber at constant volume

(A) I only

(B) II only

(C) I and II only

(D) I and III only

36. A According to Le Châtelier's law, the equilibrium will shift to counteract any stress that is placed on it.

If N_2 is added, the equilibrium will shift to remove the excess N_2. This shift produces more products, and thus, more NH_3. So (I) is correct.

Choice (II) is wrong because increasing the volume will cause the reaction to shift toward the side with more moles of gas. In this case the reactant side has more moles of gas (4 moles) than the product side (2 moles). So increasing the volume will decrease the number of moles of NH_3.

Choice (III) is wrong because increasing the temperature will favor the endothermic direction, which in this case is the reverse reaction. So once again, the number of moles of NH_3 is decreased.

37.

$$CH_4 + 2\,O_2 \rightarrow CO_2 + 2\,H_2O$$

If 16 grams of CH_4 reacts with 16 grams of O_2 in the reaction shown above, which of the following will be true?

(A) The mass of H_2O formed will be twice the mass of CO_2 formed.

(B) Equal masses of H_2O and CO_2 will be formed.

(C) Equal numbers of moles of H_2O and CO_2 will be formed.

(D) The limiting reagant will be O_2.

37. D $Moles = \dfrac{grams}{MW}$

$$Moles\ of\ CH_4 = \frac{(16\ g)}{(16\ g/mol)} = 1\ mole$$

$$Moles\ of\ O_2 = \frac{(16\ g)}{(32\ g/mol)} = 0.5\ moles$$

From the balanced equation, 2 moles of O_2 are used up for every mole of CH_4.

When all 0.5 moles of O_2 are used up, only 0.25 moles of CH_4 will be used up, so oxygen is the limiting reagant.

Choices (A), (B), and (C) are wrong because more moles of H_2O will be formed and a greater mass of CO_2 will be formed.

38. Which of the following equilibrium expressions represents the hydrolysis of the CN^- ion?

(A) $K = \dfrac{[HCN][OH^-]}{[CN^-]}$

(B) $K = \dfrac{[CN^-][OH^-]}{[HCN]}$

(C) $K = \dfrac{[CN^-][H_3O^+]}{[HCN]}$

(D) $K = \dfrac{[HCN][H_3O^+]}{[CN^-]}$

38. A The hydrolysis of the CN^- ion is shown by the reaction below.

$$CN^- + H_2O \leftrightarrow HCN + OH^-$$

Putting products over the reactants in the equilibrium expression and omitting water because it is a pure liquid, we get

$$K = \frac{[HCN][OH^-]}{[CN^-]}$$

QUESTIONS	EXPLANATIONS

39. Which of the following is true under any conditions for a reaction that is spontaneous at any temperature?

    (A) ΔG, ΔS, and ΔH are all positive.
    (B) ΔG, ΔS, and ΔH are all negative.
    (C) ΔG and ΔS are positive, and ΔH is negative.
    (D) ΔG and ΔH are negative, and ΔS is positive.

39. **D** For a spontaneous reaction, ΔG is always negative.

From the equation $\Delta G = \Delta H - T\Delta S$, we can see that the conditions that will make ΔG always negative are when ΔH is negative and ΔS is positive.

40. Which of the following pairs of compounds are isomers?

    (A) $HCOOH$ and CH_3COOH
    (B) CH_3CH_3CHO and C_3H_7OH
    (C) C_2H_5OH and CH_3OCH_3
    (D) C_2H_4 and C_2H_6

40. **C** Isomers are different molecules that have the same collection of atoms arranged in different ways. Both of the molecules in choice (C) have 2 carbons, 6 hydrogens, and 1 oxygen.

41. A sample of an ideal gas confined in a rigid 5.00 liter container has a pressure of 363 mmHg at a temperature of 25°C. Which of the following expressions will be equal to the pressure of the gas if the temperature of the container is increased to 35°C?

    (A) $\dfrac{(363)(35)}{(25)}$ mmHg

    (B) $\dfrac{(363)(25)}{(35)}$ mmHg

    (C) $\dfrac{(363)(308)}{(298)}$ mmHg

    (D) $\dfrac{(363)(298)}{(308)}$ mmHg

41. **C** From the ideal gas laws, we know that with volume constant

$$\frac{P_1}{T_1} = \frac{P_2}{T_2}$$

Solving for P_2, we get

$$P_2 = \frac{P_1 T_2}{T_1} = \frac{(363)(308)}{(298)} \text{ mmHg}$$

42. If the temperature at which a reaction takes place is increased, the rate of the reaction will

    (A) increase if the reaction is endothermic and decrease if the reaction is exothermic.
    (B) decrease if the reaction is endothermic and increase if the reaction is exothermic.
    (C) increase if the reaction is endothermic and increase if the reaction is exothermic.
    (D) decrease if the reaction is endothermic and decrease if the reaction is exothermic.

42. **C** An increase in temperature always increases the rate of a reaction, regardless of the change in enthalpy of the reaction.

Temperature change and enthalpy come into play in establishing whether reactants or products are favored at equilibrium according to Le Châtelier's law, but increasing the temperature will bring the reaction to equilibrium more quickly, regardless of whether the equilibrium favors reactants or products.

QUESTIONS	EXPLANATIONS

43. The acid dissociation constant for HClO is 3.0×10^{-8}. What is the hydrogen ion concentration in a $0.12\ M$ solution of HClO?

   (A) $3.6 \times 10^{-8}\ M$
   (B) $6.0 \times 10^{-8}\ M$
   (C) $2.0 \times 10^{-5}\ M$
   (D) $6.0 \times 10^{-5}\ M$

43. **D** Use the formula for K_a.

$$K_a = \frac{[H^+][ClO^-]}{[HClO]}$$

We're trying to figure out $[H^+]$. Since for every HClO that dissociates, you get 1 H^+ and 1 ClO^-, we can rewrite the expression with $[H^+] = [ClO^-] = x$.

$$3.0 \times 10^{-8} = \frac{x^2}{0.12}$$
$$x^2 = (0.12)(3.0 \times 10^{-8}) = 0.36 \times 10^{-8}$$
$$x = 0.60 \times 10^{-4} = 6.0 \times 10^{-5}$$

So the hydrogen ion concentration is $6.0 \times 10^{-5}\ M$.

44. $$2\,NO(g) + 2\,H_2(g) \rightarrow N_2(g) + 2\,H_2O(g)$$

   Which of the following is true regarding the relative molar rates of disappearance of the reactants and appearance of the products?

   I. N_2 appears at the same rate that H_2 disappears.
   II. H_2O appears at the same rate that NO disappears.
   III. NO disappears at the same rate that H_2 disappears.

   (A) I only
   (B) I and II only
   (C) II and III only
   (D) I, II, and III

44. **C** For every mole of N_2 that appears, 2 moles of H_2 must disappear, so N_2 appears at half the rate that H_2 disappears, and (I) is wrong.

For every 2 moles of H_2O that appear, 2 moles of NO must disappear, so H_2O appears at the same rate that NO disappears, and (II) is correct.

For every 2 moles of NO that disappear, 2 moles of H_2 must disappear, so NO disappears at the same rate that H_2 disappears, and (III) is correct.

45. $$2\,ZnS(s) + 3\,O_2(g) \rightarrow 2\,ZnO(s) + 2\,SO_2(g)$$

   If the reaction above took place at standard temperature and pressure, what was the volume of $O_2(g)$ required to produce 40.0 grams of $ZnO(s)$?

   (A) $\dfrac{(40.0)(2)}{(81.4)(3)(22.4)}\ L$
   (B) $\dfrac{(40.0)(3)}{(81.4)(2)(22.4)}\ L$
   (C) $\dfrac{(40.0)(2)(22.4)}{(81.4)(3)}\ L$
   (D) $\dfrac{(40.0)(3)(22.4)}{(81.4)(2)}\ L$

45. **D** $\text{Moles} = \dfrac{\text{grams}}{\text{MW}}$

$$\text{Moles of ZnO} = \frac{(40.0\ g)}{(81.4\ g/mol)}$$

For every 2 moles of ZnO produced, 3 moles of O_2 are consumed.

So moles of $O_2 = \left(\dfrac{3}{2}\right)(\text{moles of ZnO}) = \left(\dfrac{3}{2}\right)\dfrac{(40.0)}{(81.4)}$

At STP, volume of gas $= (\text{moles})(22.4\ L)$

So volume of $O_2 = \left(\dfrac{3}{2}\right)\dfrac{(40.0)}{(81.4)}\ (22.4)\ L$

QUESTIONS	EXPLANATIONS

46. Which of the following salts will produce a colorless solution when added to water?

(A) $Cu(NO_3)_2$
(B) $NiCl_2$
(C) $KMnO_4$
(D) $ZnSO_4$

46. D Zn^{2+} and SO_4^{2-} are both colorless in solution.

About the other answers

(A) Cu^{2+} ions are blue in solution.
(B) Ni^{2+} ions are green in solution.
(C) MnO_4^- ions are purple in solution.

Most salts of transition metals produce colored solutions. That's because energy is released and absorbed when d subshell electrons change energy levels. This energy is manifested as visible light.

Zn^{2+} does not produce a colored solution because it has a full $3d$ subshell. The full subshell means that there are no empty orbitals for the d electrons to jump to. If there are no transitions, there is no light.

47. A beaker contains 300.0 ml of a 0.20 M $Pb(NO_3)_2$ solution. If 200.0 ml of a 0.20 M solution of $MgCl_2$ is added to the beaker, what will be the final concentration of Pb^{2+} ions in the resulting solution?

(A) 0.020 M
(B) 0.040 M
(C) 0.080 M
(D) 0.120 M

47. B The Pb^{2+} ions and the Cl^- ions will combine and precipitate out of the solution, so let's find out how many of each we have. Remember that each mole of $MgCl_2$ produces 2 moles of Cl^-.

Moles = (molarity)(volume)

Moles of Pb^{2+} = (0.20 M)(0.300 L) = 0.060 mole

Moles of Cl^- = (2)(0.20 M)(0.200 L) = 0.080 mole

Since two Cl^- ions are required for each Pb^{2+} ion, subtract half the moles of Cl^- (0.040 mole) from the moles of Pb^{2+} to find the number of moles of Pb^{2+} left in the solution

0.060 mol – 0.040 mol = 0.020 mol

Now use the formula for molarity to find the concentration of Pb^{2+} ions. Don't forget to add the volumes of the two solutions.

$$\text{Molarity} = \frac{\text{moles}}{\text{liters}} = \frac{(0.020 \text{ mol})}{(0.500 \text{ L})} = 0.040 \, M$$

QUESTIONS	EXPLANATIONS

48.

$$H_2(g) + I_2(g) \rightarrow 2\,HI(g)$$

At 450°C the equilibrium constant, K_c, for the reaction shown above has a value of 50. Which of the following sets of initial conditions at 450°C will cause the reaction above to produce more H_2?

   I. [HI] = 5-molar, [H_2] = 1-molar, [I_2] = 1-molar

  II. [HI] = 10-molar, [H_2] = 1-molar, [I_2] = 1-molar

 III. [HI] = 10-molar, [H_2] = 2-molar, [I_2] = 2-molar

(A) I only
(B) II only
(C) I and II only
(D) II and III only

48. B The reaction will proceed in the reverse direction and produce more H_2 when the reaction quotient Q is greater than K_c.

Q takes the form of the equilibrium constant.

$$Q = \frac{[HI]^2}{[H_2][I_2]}$$

For I: $Q = \dfrac{(5)^2}{(1)(1)} = 25$ Q is less than K_c.

For II: $Q = \dfrac{(10)^2}{(1)(1)} = 100$ Q is greater than K_c.

For III: $Q = \dfrac{(10)^2}{(2)(2)} = 25$ Q is less than K_c.

So only (II) will produce more H_2.

49. Which of the following statements is true regarding sodium and potassium?

(A) Sodium has a larger first ionization energy and a larger atomic radius.
(B) Sodium has a larger first ionization energy and a smaller atomic radius.
(C) Sodium has a smaller first ionization energy and a larger atomic radius.
(D) Sodium has a smaller first ionization energy and a smaller atomic radius.

49. B Sodium's valence electron is in the third energy level and potassium's valence electron is in the fourth energy level, so sodium's valence electron is closer to the nucleus than potassium's, so sodium must have a smaller atomic radius.

Also, because sodium's valence electron is closer to the nucleus, it is more difficult to remove, so sodium will have a higher first ionization energy.

50.

$$HCl(aq) + AgNO_3(aq) \rightarrow AgCl(s) + HNO_3(aq)$$

One-half liter of a 0.20-molar HCl solution is mixed with one-half liter of a 0.40-molar solution of $AgNO_3$. A reaction occurs forming a precipitate as shown above. If the reaction goes to completion, what is the mass of AgCl produced?

(A) 14 grams
(B) 28 grams
(C) 42 grams
(D) 70 grams

50. A First we have to find the limiting reagent.

Moles = (molarity)(liters)

Moles of HCl = (0.20 M)(0.50 L) = 0.10 moles

Moles of $AgNO_3$ = (0.40 M)(0.50 L) = 0.20 moles

From the balanced equation, the two reactants are used up at equal rates. There is twice as much $AgNO_3$, so when the 0.10 moles of HCl have been used up, there will still be 0.10 moles of $AgNO_3$. So HCl is the limiting reagent.

From the balanced equation, for every mole of HCl consumed, 1 mole of AgCl is produced. So 0.10 moles of AgCl will be produced.

Grams = (moles)(MW)

Grams of AgCl = (0.10 moles)(143 g/mol) = 14 grams

QUESTIONS	EXPLANATIONS

51.

$$H_2(g) + Cl_2(g) \rightarrow 2\ HCl(g)$$

Based on the information given in the table below, what is ΔH for the above reaction?

Bond	Average Bond Energy (kJ/mol)
H–H	440
Cl–Cl	240
H–Cl	430

(A) −860 kJ
(B) −620 kJ
(C) −440 kJ
(D) −180 kJ

51. D The bond energy is the energy that must be put into a bond to break it.

First let's figure out how much energy must be put in to the reactants to break their bonds.

To break 1 mole of H–H bonds, it takes 440 kJ.

To break 1 mole of Cl–Cl bonds, it takes 240 kJ.

So to break up the reactants, it takes +680 kJ.

Energy is given off when a bond is formed; that's the negative of the bond energy.

Now let's see how much energy is given off when 2 moles of HCl are formed.

2 moles of HCl molecules contain 2 moles of HCl bonds, so $(2)(−430)$ kJ = −860 kJ are given off.

So the value of ΔH for the reaction is

$(−860,\ E$ given off$) + (680,\ E$ put in$) = −180$ kJ.

52. The first ionization energy for magnesium is 730 kJ/mol. The third ionization energy for magnesium is 7,700 kJ/mol. What is the most likely value for magnesium's second ionization energy?

(A)   490 kJ/mol
(B)   1,400 kJ/mol
(C)   4,200 kJ/mol
(D)   8,400 kJ/mol

52. B Magnesium has two valence electrons in the third shell, so we would expect to see a small jump between the first and second ionization energies.

The third electron must be removed from the second shell, so we would expect to see a much larger jump between the second and third ionization energies.

Choice (B) is the only answer that shows this relationship.

QUESTIONS	EXPLANATIONS

53. Molten NaCl is electrolyzed with a constant current of 1.00 ampere. What is the shortest amount of time, in seconds, that it would take to produce 1.00 mole of solid sodium? (1 faraday = 96,500 coulombs)

    (A) 32,200 seconds
    (B) 48,300 seconds
    (C) 64,300 seconds
    (D) 96,500 seconds

53. **D** First let's find out how many moles of electrons we need.

The half-reaction that reduces Na^+ to $Na(s)$ is as follows:

$Na^+ + e^- \rightarrow Na(s)$

So it takes 1 mole of electrons to produce 1 mole of $Na(s)$.

Now let's find out how many coulombs we need.

$$\text{Moles of electrons} = \frac{\text{coulombs}}{96,500}$$

So coulombs = (moles of electrons)(96,500)

    = (1)(96,500) = 96,500 coulombs.

Now we can find how many seconds it takes

$$\text{Amperes} = \frac{\text{coulombs}}{\text{second}}$$

So seconds = $\dfrac{\text{coulombs}}{\text{amperes}} = \dfrac{(96,500)}{(1)}$

    = 96,500 seconds

54. How many moles of KCl must be added to 200 milliliters of a 0.5-molar NaCl solution to create a solution in which the concentration of Cl^- ion is 1.0-molar? (Assume the volume of the solution remains constant.)

    (A) 0.1 moles
    (B) 0.2 moles
    (C) 0.3 moles
    (D) 0.4 moles

54. **A** For every NaCl in solution, there's one Cl^- ion; and for every KCl we add, we get one Cl^- ion.

Let's find out how many moles of Cl^- ions are already in the solution.

Moles = (molarity)(volume)

Moles of Cl^- = (0.5 M)(0.2 L) = 0.1 moles

We're not changing the volume, so to double the concentration of Cl^- ions from 0.5-molar to 1.0-molar, we just double the number of moles of Cl^- ions. We do that by adding 0.1 moles of KCl.

QUESTIONS	EXPLANATIONS

55. A student placed solid barium oxalate in a beaker filled with distilled water and allowed it to come to equilibrium with its dissolved ions. The student then added a nickel (II) nitrate solution to the beaker to create the following equilibrium situation.

$$BaC_2O_4 + Ni^{2+} \leftrightarrow Ba^{2+} + NiC_2O_4$$

If the solubility product for BaC_2O_4 is 2×10^{-7} and the solubility product for NiC_2O_4 is 4×10^{-10}, which of the following gives the value of the equilibrium constant for the reaction above?

(A) 2×10^3
(B) 5×10^2
(C) 5×10^{-2}
(D) 2×10^{-3}

55. B We can think of the reaction given in the question as the sum of two other reactions.

$$BaC_2O_4 \leftrightarrow Ba^{2+} + C_2O_4^{2-} \quad K_1 = 2 \times 10^{-7}$$

$$Ni^{2+} + C_2O_4^{2-} \leftrightarrow NiC_2O_4 \quad K_2 = \frac{1}{\left(4 \times 10^{-10}\right)}$$

Notice that we are using the reverse reaction for the solvation of NiC_2O_4, so the reactants and products are reversed, and we must take the reciprocal of the solubility product.

When reactions can be added to get another reaction, their equilibrium constants can be multiplied to get the equilibrium constant of the resulting reaction.

So $K_{eq} = (K_1)(K_2) = (2 \times 10^{-7}) \dfrac{1}{\left(4 \times 10^{-10}\right)}$

$= 0.5 \times 10^3 = 5 \times 10^2$

56. A 100-gram sample of pure $_{18}^{37}Ar$ decays by electron capture with a half-life of 35 days. How long will it take for 90 grams of $_{17}^{37}Cl$ to accumulate?

(A) 31 days
(B) 39 days
(C) 78 days
(D) 116 days

56. D Because they have the same mass number, the mass of $_{17}^{37}Cl$ will accumulate at the same rate that the mass of $_{18}^{37}Ar$ disappears.

We're looking for the moment when 10 grams of $_{18}^{37}Ar$ remains.

Make a chart. Start at time = 0.

Half-Lives	Time	Stuff
0	0 days	100 g
1	35 days	50 g
2	70 days	25 g
3	105 days	12.5 g
4	140 days	6.25 g

It takes between 3 and 4 half-lives for the amount of $_{18}^{37}Ar$ to decrease to 10 grams.

116 days is the only answer choice between 105 days and 140 days.

QUESTIONS	EXPLANATIONS

57. The solubility product, K_{sp}, of CaF_2 is 4×10^{-11}. Which of the following expressions is equal to the solubility of CaF_2?

(A) $\sqrt{2 \times 10^{-11}}$ M

(B) $\sqrt[3]{4 \times 10^{-11}}$ M

(C) $\sqrt[3]{2 \times 10^{-11}}$ M

(D) $\sqrt[3]{1 \times 10^{-11}}$ M

57. **D** The solubility of a substance is equal to its maximum concentration in solution.

For every CaF_2 in solution, we get one Ca^{2+} and two F^-, so the solubility of CaF_2 (let's call it x) will be the same as $[Ca^{2+}]$.

So for CaF_2

$$K_{sp} = [Ca^{2+}][F^-]^2$$
$$4 \times 10^{-11} = (x)(2x)^2 = 4x^3$$
$$x^3 = 1 \times 10^{-11}$$
$$x = \sqrt[3]{1 \times 10^{-11}}$$

58. When excess hydroxide ions were added to 1.0 liter of $CaCl_2$ solution, $Ca(OH)_2$ precipitate was formed. If all of the calcium ions in the solution were precipitated in 7.4 grams of $Ca(OH)_2$, what was the initial concentration of the $CaCl_2$ solution?

(A) 0.05-molar
(B) 0.10-molar
(C) 0.15-molar
(D) 0.20-molar

58. **B** $\text{Moles} = \dfrac{\text{grams}}{\text{MW}}$

Moles of $Ca(OH)_2 = \dfrac{(7.4 \text{ g})}{(74 \text{ g/mol})} = 0.10$ moles

One mole of $CaCl_2$ must have been consumed for every mole of $Ca(OH)_2$ produced, so there must have been 0.10 moles of $CaCl_2$ in the original solution.

$\text{Molarity} = \dfrac{\text{moles}}{\text{liters}}$

Molarity of $CaCl_2 = \dfrac{(0.10 \text{ mol})}{(1.0 \text{ L})} = 0.10$-molar

59. When a solution of $KMnO_4$ was mixed with a solution of HCl, Cl_2 gas bubbles formed and Mn^{2+} ions appeared in the solution. Which of the following has occurred?

(A) K^+ has been oxidized by Cl^-.
(B) K^+ has been oxidized by H^+.
(C) Cl^- has been oxidized by K^+.
(D) Cl^- has been oxidized by MnO_4^-.

59. **D** The oxidation and reduction half-reactions are as follows:

Cl^- is oxidized: $\quad 2\,Cl^- \rightarrow Cl_2 + 2e^-$

MnO_4^- is reduced: $\quad Mn^{7+} + 5e^- \rightarrow Mn^{2+}$

So Cl^- is oxidized and MnO_4^- is the oxidizing agent.

QUESTIONS	EXPLANATIONS

60. $2\ Cu^+(aq) + M(s) \rightarrow 2\ Cu(s) + M^{2+}(aq)$

$E^o = +0.92$ V

$Cu^+(aq) + e^- \rightarrow Cu(s)$

$E^o = +0.52$ V

Based on the reduction potentials given above, what is the standard reduction potential for the following half-reaction?

$M^{2+}(aq) + 2\ e^- \rightarrow M(s)$

(A) +0.40 V
(B) +0.12 V
(C) −0.12 V
(D) −0.40 V

60. **D** We subtract the reduction potential for Cu^+ (+0.52 V) from the reaction potential for the full reaction (+0.92 V) to get +0.40 V, the *oxidation* potential for $M(s)$. Remember: Ignore the coefficients in the reaction when you're calculating reaction potentials. But you're not done. You want the *reduction* potential for $M^{2+}(aq)$, the reverse reaction, so you need to change the sign you got for the oxidation potential. So the answer is −0.40 V.

QUESTIONS	EXPLANATIONS

SECTION II—FREE RESPONSE

1. A 0.20-molar solution of acetic acid, $HC_2H_3O_2$, at a temperature of 25°C, has a pH of 2.73.

   (a) Calculate the hydroxide ion concentration, $[OH^-]$.

   (b) What is the value of the acid ionization constant, K_a, for acetic acid at 25°C?

   (c) How many moles of sodium acetate must be added to 500.0 ml of a 0.200-molar solution of acetic acid to create a buffer with a pH of 4.00? Assume that the volume of the solution is not changed by the addition of sodium acetate.

(a) Knowing the pH, we can calculate the pOH.

$$pH + pOH = 14$$
$$2.73 + pOH = 14$$
$$pOH = 11.27$$

Knowing the pOH, we can calculate $[OH^-]$

$$[OH^-] = 10^{-pOH} = 10^{-11.27} = 5.4 \times 10^{-12}$$

(b)
$$K_a = \frac{[H^+][C_2H_3O_2^-]}{[HC_2H_3O_2]} = \frac{x^2}{0.200 - x}$$

Knowing the pH, we can find $[H^+]$. We also know that every $HC_2H_3O_2$ molecule that dissociates will put 1 H^+ ion and 1 $C_2H_3O_2$ ion in solution.

$$x = [H^+] = [C_2H_3O_2^-] = 10^{-pH} = 10^{-2.73} = 1.86 \times 10^{-3}$$

$$[HC_2H_3O_2] = 0.200 - x$$

x is very small, so $[HC_2H_3O_2] = 0.200$

Now we can solve for K_a.

$$K_a = \frac{[H^+][C_2H_3O_2^-]}{[HC_2H_3O_2]} = \frac{x^2}{0.200} = \frac{(1.86 \times 10^{-3})^2}{(0.200)}$$

$$= 1.7 \times 10^{-5}$$

(c) We can use the Henderson-Hasselbalch expression to find out what value of $[C_2H_3O_2]$ will create a buffer with a pH of 4.

$$pH = pK_a + \log \frac{[A^-]}{[HA]}$$

$$pH = pK_a + \log \frac{[C_2H_3O_2^-]}{[HC_2H_3O_2]}$$

$$pH = 4.00$$
$$pK_a = -\log(1.73 \times 10^{-5}) = 4.76$$

$$\log \frac{[C_2H_3O_2^-]}{[HC_2H_3O_2]} = pH - pK_a = 4.00 - 4.76 = -0.76$$

$$\frac{[C_2H_3O_2^-]}{[HC_2H_3O_2]} = 10^{-0.76} = 0.174$$

$$[HC_2H_3O_2] = 0.200 \, M$$
$$[C_2H_3O_2^-] = (0.174)(0.200) = 0.035 \, M$$

Moles = (molarity)(volume) = (0.035)(0.500)
$$= 0.018 \text{ moles}$$

QUESTIONS	EXPLANATIONS

(d) In a titration experiment, 100.0 ml of sodium hydroxide solution was added to 200 ml of a 0.400-molar solution of acetic acid until the equivalence point was reached. What was the pH at the equivalence point?

(d) Because $HC_2H_3O_2$ dissociates to such a small extent, we can assume that all of the $C_2H_3O_2^-$ in the solution came from the NaC_2H_3O.

Use the base ionization constant for $C_2H_3O_2^-$.

$$K_b = \frac{[HC_2H_3O_2][OH^-]}{[C_2H_3O_2^-]}$$

$$K_b = \frac{10^{-14}}{K_a} = \frac{10^{-14}}{1.73 \times 10^{-5}} = 5.78 \times 10^{-10}$$

At the equivalence point, all of the acetic acid initially present has been converted to acetate ion. So the initial $[HC_2H_3O_2]$ is equal to $[C_2H_3O_2]$ at the equivalence point.

Moles = (molarity)(volume)

Moles of $C_2H_3O_2^- = (0.400\ M)(0.200\ L)$
$= 0.080$ moles

$$Molarity = \frac{moles}{volume}$$

$$[C_2H_3O_2^-] = \frac{(0.080\ mol)}{(0.200\,L + 0.100\,L)}$$

$$= \frac{(0.080\ mol)}{(0.300\,L)} = 0.267\ M$$

$[HC_2H_3O_2] = [OH^-] = x$

Now we can use the K_b equation to find x, the OH^- concentration.

We'll assume that x is much smaller than $0.267\ M$.

$$K_b = \frac{[HC_2H_3O_2][OH^-]}{[C_2H_3O_2^-]} = \frac{x^2}{0.267 - x} = \frac{x^2}{0.267}$$

$$K_b = 5.78 \times 10^{-10} = \frac{x^2}{0.267}$$
$x = [OH^-] = 1.24 \times 10^{-5}\ M$

Knowing $[OH^-]$, we can calculate pOH, and then pH.

pOH $= -\log[OH^-] = -\log(1.24 \times 10^{-5}) = 4.91$

pH $= 14 - $ pOH

pH $= 14 - 4.91 = 9.09$

QUESTIONS	EXPLANATIONS

(e) The pK_a values for several indicators are given in the table below. Which of the indicators on the table is most suitable for this titration? Justify your answer.

Indicator	pK_a
Thymol Blue	2
Bromcresol Purple	6
Phenolphthalein	9

(e) Phenolphthalein, with a pK_a of 9, is the best choice. The pH at the equivalence point is about 9, and the indicator should have a pK_a that is close to the equivalence point for the titration.

2.
$$2 \, NO(g) + Cl_2(g) \rightarrow 2 \, NOCl(g)$$

The following data were collected for the reaction above. All of the measurements were taken at a temperature of 263 K.

Experiment	Initial [NO] (M)	Initial [Cl$_2$] (M)	Initial rate of disappearance of Cl$_2$ (M/min)
1	0.15	0.15	0.60
2	0.15	0.30	1.2
3	0.30	0.15	2.4
4	0.25	0.25	?

(a) Write the expression for the rate law for the reaction above.

(a) From experiments 1 and 2, we can see that when [Cl$_2$] doubles, the rate doubles, so the rate law is first order with respect to Cl$_2$. That is, $[Cl_2]^1$.

From experiments 1 and 3, we can see that when [NO] doubles, the rate quadruples, so the rate law is second order with respect to NO. That is, $[NO]^2$.

Rate $= k[NO]^2[Cl_2]$

(b) Calculate the value of the rate constant for the above reaction, and specify the units.

(b) We'll use experiment 1 for our calculation.

$$k = \frac{Rate}{[NO]^2[Cl_2]} = \frac{(0.60 \, M/min)}{(0.15 \, M)^2 (0.15 \, M)} = 180 \, M^{-2} \, min^{-1}$$

(c) What is the initial rate of appearance of NOCl in experiment 2?

(c) From the balanced equation we can see that for every molecule of Cl$_2$ that disappears, 2 molecules of NOCl appear, so the rate of appearance of NOCl will be twice the rate of disappearance of Cl$_2$.

In experiment 2, the initial rate of disappearance of Cl$_2$ is 1.2 M/min, so the initial rate of appearance of NOCl will be 2.4 M/min.

QUESTIONS	EXPLANATIONS

(d) What is the initial rate of disappearance of Cl_2 in experiment 4?

(d) Use the rate law.

Rate = $k[NO]^2[Cl_2]$

Rate = $(180\ M^{-2}\ min^{-1})(0.25\ M)^2(0.25\ M)$
    $= 2.8\ M/min$

(e) Each of the experimental trials took place in a closed container. Explain or calculate each of the following:
(i) What was the partial pressure due to $NO(g)$ at the start of experiment 1?
(ii) What was the total pressure at the start of experiment 1? Assume that no NOCl is present.

(e) (i) Use the gas law

$$P = \frac{nRT}{V} = MRT$$

$P = (0.15\ mol/L)(0.0821\ L\text{-}atm/mol\text{-}K)(263K) = 3.2$ atm

(ii) Since the initial concentrations of NO and Cl_2 are the same for experiment 1, the initial partial pressure due to Cl_2 will also be 3.2 atm. From Dalton's law, the total pressure is equal to the sum of the partial pressures. So 3.2 atm + 3.2 atm = 6.4 atm.

3. A titration experiment was conducted to determine the pH of a known volume of a strong monoprotic acid solution. A hydroxide solution of known concentration was poured into a buret and then titrated into the acid solution. The volume of hydroxide solution titrated into the acid solution was measured at the equivalence point and used to calculate the concentration of the acid solution. Which of the following situations would cause an error in the calculated value of the pH of the acid solution? Explain each answer.

In this experiment, the volume of OH^- added is directly measured.

Moles = (molarity)(volume) is used to find the moles of OH^- added.

The moles of OH^- added to reach the equivalence point is equal to the number of moles of H^+ originally present in the acid solution.

Molarity = $\dfrac{moles}{volume}$ is used to find $[H^+]$ of the acid solution.

$pH = -\log[H^+]$ is used to calculate the pH.

(a) The buret was rinsed with the hydroxide solution before the solution was poured into the buret.

(a) This is proper experimental procedure and will have no adverse effect on the calculated value of the acid solution.

(b) The experimenter did not notice that a few drops of hydroxide solution spattered outside the acid solution container during the titration.

(b) This will make the measured volume of the hydroxide solution larger than the actual amount added, the calculated value for the moles of OH^- and H^+ will be too large, the calculated value of $[H^+]$ will be too large, and the calculated pH will be too small.

QUESTIONS	EXPLANATIONS

(c) The buret was rinsed with distilled water before the hydroxide solution was added.

(c) This will dilute the hydroxide solution, which means that too large a volume of hydroxide solution will be added, the calculated value for the moles of OH^- and H^+ will be too large, the calculated $[H^+]$ will be too large, and the calculated pH will be too small.

(d) The experimenter read the hydroxide solution level from the top of the fluid instead of the bottom of the meniscus both before the titration and at the equivalence point.

(d) The levels were read consistently although they were read from the wrong spot. The two errors should cancel, and the calculated pH should be correct.

(e) Some hydroxide solution was spilled while the experimenter was pouring it into the buret.

(e) This will not affect the concentration of the hydroxide solution or the measurement of the volume poured into the acid solution, so the calculated pH should not be affected.

4.
$$CH_4(g) + 2\,O_2(g) \rightarrow CO_2(g) + 2\,H_2O(l)$$

The above reaction for the combustion of methane gas has a standard entropy change, ΔS°, with a value of -242.7 J/mol-K. The following data are also available:

Compound	ΔH_f (kJ/mol)
$CH_4\,(g)$	-74.8
$H_2O\,(l)$	-285.9
$CO_2\,(g)$	-393.5

(a) What are the values of ΔH_f° and ΔG_f° for $O_2(g)$?

(a) ΔH_f° and ΔG_f° for $O_2(g)$ are both equal to zero. The enthalpy and free energy of formation of any element in its standard state are equal to zero.

(b) Calculate the standard change in enthalpy, ΔH°, for the combustion of methane.

(b) $\Delta H = \Sigma\,\Delta H_f^\circ\,(\text{products}) - \Sigma\,\Delta H_f^\circ\,(\text{reactants})$

$\Delta H = [(-393.5) + (2)(-285.9)]\text{ kJ} - [-74.8]\text{ kJ}$

$\Delta H = [-965.3]\text{ kJ} - [-74.8]\text{ kJ}$

$\Delta H = -890.5\text{ kJ}$

QUESTIONS	EXPLANATIONS
(c) Calculate the standard free energy change, $\Delta G°$, for the combustion of methane.	(c) $\Delta G° = \Delta H° - T\,\Delta S°$ $\Delta G° = (-890{,}500 \text{ J}) - (298 \text{ K})(-242.7 \text{ J/K})$ $\Delta G° = -818{,}200 \text{ J} = -818.2 \text{ kJ}$
(d) How would the value of $\Delta S°$ for the reaction be affected if the water produced in the combustion remained in the gas phase?	(d) $\Delta S°$ would become less negative. $H_2O(g)$ has more entropy than $H_2O(l)$, so the entropy of the products would be increased and the entropy change of the reaction would become more positive (less negative, that is).
(e) A 20.0-gram sample of $CH_4(g)$ underwent combustion in a bomb calorimeter with excess oxygen gas. (i) Calculate the mass of carbon dioxide produced. (ii) Calculate the heat released by the reaction.	(e) (i) First find moles of CH_4. $\text{Moles} = \dfrac{\text{grams}}{\text{MW}} = \dfrac{20.0 \text{ grams}}{16.0 \text{ g/mol}} = 1.25 \text{ mol}$ There is a one to one ratio in the balance equation between CH_4 and CO_2, so 1.25 moles of CO_2 are produced. Now find the grams of CO_2. $\text{Grams} = (\text{moles})(\text{MW}) = (1.25 \text{ mol})(44.0 \text{ g/mol}) = 55$ grams (ii) From (c), the change in enthalpy when one mole of CH_4 is consumed is 890.5 kJ. So when 1.25 moles are consumed, $\Delta H = (890.5 \text{ kJ/mol})(1.25 \text{ mol}) = 1{,}110 \text{ kJ}$. So about 1,110 kJ were released.

QUESTIONS	EXPLANATIONS

5. You will be given three chemical reactions below. In part (i), write the balanced equation for the reaction, leaving coefficients in terms of lowest whole numbers. Then answer the question pertaining to that reaction in part (ii). For each of the following three reactions, assume that solutions are aqueous unless it says otherwise. Substances in solutions should be represented as ions if these substances are extensively ionized. Omit formulas for ions or molecules that are not affected by the reaction. Only equations inside the answer boxes will be graded.

(a)   Sulfur dioxide gas is bubbled through cold water.

(i)   Balanced equation:

(i) $SO_2 + H_2O \rightarrow H_2SO_3$

(ii) As the reaction progresses, will the hydroxide concentration in the solution increase, decrease, or remain the same?

(ii) Sulfur dioxide is an acid anhydride, so the hydroxide ion concentration will decrease as the solution becomes more acidic.

(b)   Ethane is burned in air.

(i)   Balanced equation:

(i) $2\,C_2H_6 + 7\,O_2 \rightarrow 4\,CO_2 + 6\,H_2O$

(ii) If the enthalpy change for the reaction was measured to be −3,100 kJ and the total heat of formation for a,ll products formed is −3,300 kJ, what is the approximate heat of formation of ethane measured in kJ/mole?

(ii) The enthalpy change for the reaction comes from subtracting the heat of formation of the reactants from the heat of formation of the products, so the heat of formation of the reactants is −3,300 kJ − (−3,100 kJ) = −200 kJ. Because the heat of formation for oxygen gas is zero and there are 2 moles of ethane in the reaction, the heat of formation of ethane is approximately −100 kJ mole.

(c)   Chlorine gas is bubbled through a solution of sodium bromide.

(i)   Balanced equation:

(i) $Cl_2 + 2\,Br^- \rightarrow 2\,Cl^- + Br_2$

(ii) Which substance is reduced in this reaction?

(ii) Chlorine gas gains electrons, so it is reduced.

QUESTIONS	EXPLANATIONS

6. Oxygen is found in the atmosphere as a diatomic gas, O_2, and as ozone, O_3. Ozone has a dipole of 0.5 debye. Oxygen has a dipole moment of zero.

(a) Draw the Lewis dot structures for both molecules.

(a)

Lewis dot structure for O_2

Lewis dot structure for O_3

(b) Use the principles of bonding and molecular structure to account for the fact that ozone has a higher boiling point than diatomic oxygen.

(b) O_3 is a polar molecule, while O_2 is nonpolar, so the dipole–dipole attractions between O_3 molecules are stronger than the van der Waals forces between O_2 molecules.

If you didn't recognize the polarity of O_3, you may have gotten some partial credit for noting that O_3 has more electrons than O_2, so O_3 will have stronger van der Waals forces between its molecules.

(c) Use the principles of bonding and molecular structure to account for the fact that ozone is more soluble than diatomic oxygen in water.

(c) Water molecules are polar. O_3 is a polar molecule, while O_2 is nonpolar. So water molecules will be more strongly attracted to O_3 molecules than they are to the nonpolar O_2 molecules.

(d) Explain why the two bonds in O_3 are of equal length and are longer than the bond length of the bond in diatomic oxygen.

(d) The bond in O_2 is a double bond. Ozone, however, has two resonance forms, each with a single and a double bond, so the two bonds in O_3 are each somewhere between a single and a double bond on average. The double bond in O_2 is stronger and shorter than the single/double resonance bonds in O_3.

(e) Elemental oxygen is strongly affected by a magnetic field. Explain why.

(e) Oxygen has unpaired electrons in the $2p$ subshell. This makes oxygen paramagnetic.

(f) For the equilibrium reaction below at 25°C and 1 atm, $K_{eq} = 10^{-57}$.

$$3\ O_2(g) \leftrightarrow 2\ O_3(g)$$

Which form of oxygen is more abundant under normal conditions? Explain.

(f) The value of the equilibrium constant is very small. This means that the reactants will be far more abundant than the products.

QUESTIONS	EXPLANATIONS

7. $NaC_2H_3O_2$, $Ba(NO_3)_2$, KCl

(a) Aqueous solutions of equal concentration of the three compounds listed above are prepared. What would an experimenter expect to observe when each of the following procedures is performed on each of the solutions?

(i) The pH of each solution is measured.

   (i) The $NaC_2H_3O_2$ solution will be slightly basic. The other two solutions will be neutral.

$NaC_2H_3O_2$ is a salt composed of the conjugate of a strong base and the conjugate of a weak acid, so it will create a basic solution.

$Ba(NO_3)_2$ and KCl are salts composed of conjugates of strong acids and bases, so they will create neutral solutions.

(ii) SO_4^{2-} ions are introduced into each solution.

   (ii) A precipitate will form in the $Ba(NO_3)_2$ solution. The other two solutions will show no change.

$BaSO_4$ is insoluble. Na_2SO_4 and K_2SO_4 are both soluble.

(iii) The freezing point of each solution is measured and the three temperatures are compared.

   (iii) The freezing points of the $NaC_2H_3O_2$ and KCl solutions will be less than 0 C and about the same. The freezing point of the $Ba(NO_3)_2$ solution will be lower than the other two.

Freezing-point depression is a colligative property, so it will depend only on the number of particles in solution, not on their identity.

$NaC_2H_3O_2$ and KCl each dissociate into two ions per molecule, while $Ba(NO_3)_2$ dissociates into three ions per unit.

Since the concentrations of all three solutions are the same, the $Ba(NO_3)_2$ solution will have the greatest freezing-point depression because it dissociates into the greatest number of particles. By the same reasoning, the freezing points of the $NaC_2H_3O_2$ and KCl solutions will be about the same.

(iv) Each solution is subjected to a flame test.

   (iv) Each flame will have a distinct color: Na^+ will be yellow, Ba^{2+} will be green, and K^+ will be purple.

QUESTIONS	**EXPLANATIONS**

(b) At 25°C and 1.0 atmosphere pressure, a balloon contains a mixture of four ideal gases: oxygen, nitrogen, carbon dioxide, and helium. The partial pressure due to each gas is 0.25 atmosphere. Use the ideas of kinetic molecular theory to answer each of the following questions:

(i) Rank the gases in increasing order of average molecular velocity and explain.

(i) CO_2, O_2, N_2, He

All of the gases have the same average kinetic energy, and $KE = \frac{1}{2}mv^2$, so the larger the molecular weight, the smaller the average velocity. This relationship is given directly by the expression

$$u_{rms} = \sqrt{\frac{3RT}{(MW)}}\ .$$

(ii) How would the volume of the balloon be affected if the temperature of the gases in the balloon were increased at constant pressure?

(ii) Volume increases.

$PV = nRT$. From the ideal gas law, we can see that when T increases and P and n are held constant, V must increase to maintain equality.

(iii) What changes to the temperature and pressure of the gases would cause deviation from ideal behavior, and which gas would be most affected?

(iii) Decreased temperature and increased pressure would eventually cause deviation from ideal behavior. Carbon dioxide would be most affected.

Deviation from ideal behavior occurs when gas molecules are brought very close together. When gas molecules are packed close together, the weak attractive forces between them become important. Also, in this situation, the volumes occupied by the individual gas molecules can no longer be ignored. Gas molecules are brought close together by low temperatures and high pressures.

Carbon dioxide would be the most affected of the four gases because it is the largest molecule and it has the most electrons. Because of this, the van der Waals forces among CO_2 molecules will be stronger than for the other gases.

Chapter 12
Practice Test 2

AP® Chemistry Exam

SECTION I: Multiple-Choice Questions

DO NOT OPEN THIS BOOKLET UNTIL YOU ARE TOLD TO DO SO.

At a Glance

Total Time
1 hour and 30 minutes
Number of Questions
60
Percent of Total Grade
50%
Writing Instrument
Pencil required

Instructions

Section I of this examination contains 60 multiple-choice questions. Fill in only the ovals for numbers 1 through 60 on your answer sheet.

CALCULATORS MAY NOT BE USED IN THIS PART OF THE EXAMINATION.

Indicate all of your answers to the multiple-choice questions on the answer sheet. No credit will be given for anything written in this exam booklet, but you may use the booklet for notes or scratch work. After you have decided which of the suggested answers is best, completely fill in the corresponding oval on the answer sheet. Give only one answer to each question. If you change an answer, be sure that the previous mark is erased completely. Here is a sample question and answer.

Sample Question   Sample Answer

Chicago is a   Ⓐ ● Ⓒ Ⓓ
(A) state
(B) city
(C) country
(D) continent

Use your time effectively, working as quickly as you can without losing accuracy. Do not spend too much time on any one question. Go on to other questions and come back to the ones you have not answered if you have time. It is not expected that everyone will know the answers to all the multiple-choice questions.

About Guessing

Many candidates wonder whether or not to guess the answers to questions about which they are not certain. Multiple choice scores are based on the number of questions answered correctly. Points are not deducted for incorrect answers, and no points are awarded for unanswered questions. Because points are not deducted for incorrect answers, you are encouraged to answer all multiple-choice questions. On any questions you do not know the answer to, you should eliminate as many choices as you can, and then select the best answer among the remaining choices.

Disclaimer

This test is an approximation of the test that you will take and some of the question types may change. For up-to-date information, please remember to check the AP Central website.

GO ON TO THE NEXT PAGE.

CHEMISTRY
SECTION I

Time—1 hour and 30 minutes

Material in the following table may be useful in answering the questions in this section of the examination.

PERIODIC CHART OF THE ELEMENTS

1 H 1.0																	2 He 4.0
3 Li 6.9	4 Be 9.0											5 B 10.8	6 C 12.0	7 N 14.0	8 O 16.0	9 F 19.0	10 Ne 20.2
11 Na 23.0	12 Mg 24.3											13 Al 27.0	14 Si 28.1	15 P 31.0	16 S 32.1	17 Cl 35.5	18 Ar 39.9
19 K 39.1	20 Ca 40.1	21 Sc 45.0	22 Ti 47.9	23 V 50.9	24 Cr 52.0	25 Mn 54.9	26 Fe 55.8	27 Co 58.9	28 Ni 58.7	29 Cu 63.5	30 Zn 65.4	31 Ga 69.7	32 Ge 72.6	33 As 74.9	34 Se 79.0	35 Br 79.9	36 Kr 83.8
37 Rb 85.5	38 Sr 87.6	39 Y 88.9	40 Zr 91.2	41 Nb 92.9	42 Mo 95.9	43 Tc (98)	44 Ru 101.1	45 Rh 102.9	46 Pd 106.4	47 Ag 107.9	48 Cd 112.4	49 In 114.8	50 Sn 118.7	51 Sb 121.8	52 Te 127.6	53 I 126.9	54 Xe 131.3
55 Cs 132.9	56 Ba 137.3	57 *La 138.9	72 Hf 178.5	73 Ta 180.9	74 W 183.9	75 Re 186.2	76 Os 190.2	77 Ir 192.2	78 Pt 195.1	79 Au 197.0	80 Hg 200.6	81 Tl 204.4	82 Pb 207.2	83 Bi 209.0	84 Po (209)	85 At (210)	86 Rn (222)
87 Fr (223)	88 Ra 226.0	89 †Ac 227.0															

*Lanthanum Series

58 Ce 140.1	59 Pr 140.9	60 Nd 144.2	61 Pm (145)	62 Sm 150.4	63 Eu 152.0	64 Gd 157.3	65 Tb 158.9	66 Dy 162.5	67 Ho 164.9	68 Er 167.3	69 Tm 168.9	70 Yb 173.0	71 Lu 175.0

†Actinium Series

90 Th 232.0	91 Pa 231.0	92 U 238.0	93 Np 237.0	94 Pu (244)	95 Am (243)	96 Cm (247)	97 Bk (247)	98 Cf (251)	99 Es (252)	100 Fm (258)	101 Md (258)	102 No (259)	103 Lr (260)

DO NOT DETACH FROM BOOK.

GO ON TO THE NEXT PAGE.

Note: For all questions involving solutions and/or chemical equations, assume that the system is in pure water and at room temperature unless otherwise stated.

Part A

Directions: Each set of lettered choices below refers to the numbered questions or statements immediately following it. Select the one lettered choice that best answers each question or best fits each statement and fill in the corresponding oval on the answer sheet. A choice may be used once, more than once, or not at all in each set.

Questions 1–3 refer to the following elements.

    (A) Ne
    (B) Li
    (C) Al
    (D) Cl

1. Which of the elements above is most commonly found as a negatively charged ion?

2. Which of the elements above has the same electron structure as Mg^{2+}?

3. Which of the elements above is found as a diatomic gas in its uncombined state?

Questions 4–7 refer to the following substances.

    (A) H_2O
    (B) NH_3
    (C) CH_4
    (D) HF

4. This molecule is nonpolar.

5. This substance forms an aqueous solution with a pH that's less than 7.

6. This molecule has a trigonal pyramidal shape.

7. This substance does NOT exhibit hydrogen bonding.

Questions 8–9 refer to an experiment in which five individual 1-liter aqueous solutions, each containing a 1-mole sample of one of the salts listed below, were subjected to various tests at room temperature.

    (A) $NaC_2H_3O_2$
    (B) NaCl
    (C) $MgBr_2$
    (D) $HC_2H_3O_2$

8. The solution containing this salt had the lowest conductivity.

9. The solution containing this salt had the highest pH.

GO ON TO THE NEXT PAGE.

Part B

Directions: Each of the questions or incomplete statements below is followed by four suggested answers or completions. Select the one that is best in each case and then fill in the corresponding oval on the answer sheet.

10. Titanium metal is prepared by heating rutile, an oxide of titanium, along with carbon and chlorine gas. By mass, rutile is 40% oxygen and 60% titanium. What is the empirical formula of rutile?

    (A) TiO
    (B) Ti_2O
    (C) TiO_2
    (D) Ti_2O_3

11. What is the weight of $NaNO_3$ (molecular weight 85.0) present in 100.0 ml of a 4.00-molar solution?

    (A) 8.50 grams
    (B) 17.0 grams
    (C) 25.5 grams
    (D) 34.0 grams

12. A mixture of gases contains 5.0 moles of oxygen, 12 moles of nitrogen, and 4.0 moles of carbon dioxide. If the partial pressure due to carbon dioxide is 1.6 atmospheres, what is the partial pressure due to oxygen?

    (A) 2.0 atm
    (B) 4.8 atm
    (C) 6.0 atm
    (D) 6.8 atm

13. Which of the diagrams below represents the enthalpy change for an endothermic reaction?

14. $$2\ NO(g) + Br_2(g) \rightarrow 2\ NOBr(g)$$

    For the reaction above, the experimental rate law is given as follows:

    $$Rate = k[NO]^2[Br_2]$$

    Which of the statements below is true regarding this reaction?

    (A) The reaction is first-order overall.
    (B) The reaction is first-order with respect to Br_2.
    (C) The reaction is first-order with respect to NO.
    (D) The reaction is second-order overall.

15. Which of the following is the most likely electron configuration of a sulfur atom in its ground state?

    (A) $1s^2\, 2s^2\, 2p^6$
    (B) $1s^2\, 2s^2\, 2p^6\, 3s^2\, 3p^2$
    (C) $1s^2\, 2s^2\, 2p^6\, 3s^2\, 3p^4$
    (D) $1s^2\, 2s^2\, 2p^6\, 3s^2\, 3p^6$

16. How many grams of carbon are present in 270 grams of glucose, $C_6H_{12}O_6$?

    (A) 12.0 grams
    (B) 18.0 grams
    (C) 72.0 grams
    (D) 108 grams

17. Calcium carbonate dissolves in acidic solutions as shown in the equation below.

    $$CaCO_3(s) + 2\ H^+(aq) \rightarrow Ca^{2+}(aq) + H_2O(l) + CO_2(g)$$

    If excess $CaCO_3$ is added to 0.250 liters of a 2.00-molar HNO_3 solution, what is the maximum volume of CO_2 gas that could be produced at standard temperature and pressure?

    (A) 5.60 liters
    (B) 11.2 liters
    (C) 16.8 liters
    (D) 33.6 liters

(A)

(B)

(C)

(D)

GO ON TO THE NEXT PAGE.

18. A monoprotic acid was titrated with a solution of NaOH. For 55.0 milliliters of the acid, 37.0 milliliters of a 0.450-molar solution of NaOH was required to reach the equivalence point. Which of the following expressions is equal to the initial concentration of the monoprotic acid?

    (A) $\dfrac{(0.450)(0.037)}{(0.055)} M$

    (B) $\dfrac{(0.450)(0.055)}{(0.037)} M$

    (C) $\dfrac{(0.055)}{(0.450)(0.037)} M$

    (D) $\dfrac{(0.037)}{(0.450)(0.055)} M$

19. $Zn(s) + NO_3^-(aq) + 10\ H^+(aq) \rightarrow$
    $$Zn^{2+}(aq) + NH_4^+(aq) + 3\ H_2O(l)$$

    Which of the following statements regarding the reaction shown above is correct?

    (A) The oxidation number of hydrogen changes from +1 to 0.
    (B) The oxidation number of hydrogen changes from +1 to −1.
    (C) The oxidation number of nitrogen changes from +5 to −3.
    (D) The oxidation number of nitrogen changes from +5 to +3.

20. $HC_2H_3O_2(aq) + ClO^-(aq) \leftrightarrow HClO(aq) + C_2H_3O_2^-(aq)$

    The standard free energy change for this reaction has a negative value. Based on this information, which of the following statements is true?

    (A) K_a for $HC_2H_3O_2(aq)$ is less than K_a for $HClO(aq)$.
    (B) K_b for $C_2H_3O_2^-(aq)$ is less than K_b for $ClO^-(aq)$.
    (C) K_{eq} for the reaction is less than 1.
    (D) The reaction occurs in the presence of a catalyst.

21. A student added 0.20 mol of NaI and 0.40 mol of KI to 3 liters of water to create an aqueous solution. What is the minimum number of moles of $Pb(C_2H_3O_2)_2$ that the student must add to the solution to precipitate out all of the I^- ions as PbI_2?

    (A) 2.40
    (B) 1.20
    (C) 0.60
    (D) 0.30

22. Which of the following statements regarding fluorine and nitrogen is NOT true?

    (A) Fluorine has greater electronegativity.
    (B) Fluorine has a greater first ionization energy.
    (C) Fluorine has a greater atomic weight.
    (D) Fluorine has a greater atomic radius.

23. Which of the statements below regarding elemental nitrogen is NOT true?

    (A) It contains one sigma bond.
    (B) It contains two pi bonds.
    (C) It has a bond order of 3.
    (D) It has a large dipole moment.

24. The reaction of elemental chlorine with ozone in the atmosphere occurs by the two-step process shown below.

    I. $Cl + O_3 \rightarrow ClO + O_2$
    II. $ClO + O \rightarrow Cl + O_2$

    Which of the statements below is true regarding this process?

    (A) Cl is a catalyst.
    (B) O_3 is a catalyst.
    (C) ClO is a catalyst.
    (D) O_2 is an intermediate.

25. A student examined the line spectrum of a hydrogen atom and was able to conclude that when the electron in the hydrogen atom made the transition from the $n = 3$ to the $n = 1$ state, the frequency of the ultraviolet radiation emitted was 2.92×10^{15} Hz. When the same transition occurred in a He^+ ion, the frequency emitted was 1.17×10^{16} Hz. Which of the following best accounts for the difference?

    (A) More energy is released by the He^+ transition because of the greater molar mass of the He^+ ion.
    (B) More energy is released by the He^+ transition because of the greater nuclear charge of the He^+ ion.
    (C) More energy is released by the He^+ transition because the ionic radius of He^+ is greater than the atomic radius of H.
    (D) Less energy is released by the He^+ transition because the ionic radius of He^+ is smaller than the atomic radius of H.

GO ON TO THE NEXT PAGE.

26. $CH_3CH_2OH(g) + ...O_2(g) \rightarrow ...CO_2(g) + ...H_2O(g)$

    The reaction above represents the oxidation of ethanol. How many moles of O_2 are required to oxidize 1 mole of CH_3CH_2OH?

    (A) $\frac{3}{2}$ moles

    (B) $\frac{5}{2}$ moles

    (C) 3 moles

    (D) $\frac{7}{2}$ moles

27. How many milliliters of water must be added to 10 milliliters of an HCl solution with a pH of 1 to produce a solution with a pH of 2?

    (A) 10 ml
    (B) 90 ml
    (C) 100 ml
    (D) 990 ml

28. Elemental iodine (I_2) is more soluble in carbon tetrachloride (CCl_4) than it is in water (H_2O). Which of the following statements is the best explanation for this?

    (A) I_2 is closer in molecular weight to CCl_4 than it is to H_2O.
    (B) The freezing point of I_2 is closer to that of CCl_4 than it is to that of H_2O.
    (C) I_2 and CCl_4 are nonpolar molecules, while H_2O is a polar molecule.
    (D) The heat of formation of I_2 is closer to that of CCl_4 than it is to that of H_2O.

29. A sample of water was electrolyzed to produce hydrogen and oxygen gas, as shown in the reaction below.

    $$2 H_2O(l) \rightarrow 2 H_2(g) + O_2(g)$$

    If 33.6 liters of gas were produced at STP, how many grams of water were consumed in the reaction?

    (A) 11.1 grams
    (B) 18.0 grams
    (C) 24.0 grams
    (D) 36.0 grams

30. The following data were gathered in an experiment to determine the density of a sample of an unknown substance.

    Mass of the sample = 7.50 grams

    Volume of the sample = 2.5 milliliters

    The density of the sample should be reported as

    (A) 3.00 grams per ml.
    (B) 3.0 grams per ml.
    (C) 3 grams per ml.
    (D) 0.3 grams per ml.

31. $$2 H_2O_2(g) + S(s) \rightarrow SO_2(g) + 2 H_2O(g)$$

    Based on the information given in the table below, what is the enthalpy change in the reaction represented above?

Substance	ΔH_f (kJ/mol)
$H_2O_2(g)$	−150
$SO_2(g)$	−300
$H_2O(g)$	−250
$S(s)$	0

    (A) −500 kJ
    (B) −200 kJ
    (C) 200 kJ
    (D) 400 kJ

32. A chemist creates a buffer solution by mixing equal volumes of a 0.2-molar HOCl solution and a 0.2-molar KOCl solution. Which of the following will occur when a small amount of KOH is added to the solution?

    I.   The concentration of undissociated HOCl will increase.
    II. The concentration of OCl⁻ ions will increase.
    III. The concentration of H⁺ ions will increase.

    (A) I only
    (B) II only
    (C) III only
    (D) I and III only

GO ON TO THE NEXT PAGE.

33. A 15-gram sample of neon is placed in a sealed container with a constant volume of 8.5 liters. If the temperature of the container is 27 C, which of the following expressions gives the correct pressure of the gas? The ideal gas constant, R, is 0.08 (L-atm)/(mole-K).

(A) $\dfrac{(15)(0.08)(300)}{(20)(8.5)}$ atm

(B) $\dfrac{(20)(0.08)(300)}{(15)(8.5)}$ atm

(C) $\dfrac{(20)(300)}{(15)(8.5)(0.08)}$ atm

(D) $\dfrac{(15)(300)}{(20)(8.5)(0.08)}$ atm

34. $$C(s) + CO_2(g) + energy \leftrightarrow 2\,CO(g)$$

The system above is currently at equilibrium in a closed container. Which of the following changes to the system would serve to increase the number of moles of CO present at equilibrium?

   I. Raising the temperature
   II. Increasing the volume of the container
   III. Adding more $C(s)$ to the container

(A) I only
(B) II only
(C) I and II only
(D) I and III only

35. In a titration experiment, a 0.10-molar $H_2C_2O_4$ solution was completely neutralized by the addition of a 0.10-molar NaOH solution. Which of the diagrams below illustrates the change in pH that accompanied this process?

(A)

pH   NaOH added

(C)

pH   NaOH added

(B)

pH   NaOH added

(D)

pH   NaOH added

36. If 52 grams of $Ba(NO_3)_2$ (molar mass 260 grams) are completely dissolved in 500 milliliters of distilled water, what are the concentrations of the barium and nitrate ions?

(A) $[Ba^{2+}] = 0.10\ M$ and $[NO_3^-] = 0.10\ M$
(B) $[Ba^{2+}] = 0.20\ M$ and $[NO_3^-] = 0.40\ M$
(C) $[Ba^{2+}] = 0.40\ M$ and $[NO_3^-] = 0.40\ M$
(D) $[Ba^{2+}] = 0.40\ M$ and $[NO_3^-] = 0.80\ M$

37. All of the following statements concerning the alkali metals are true EXCEPT

(A) They are strong oxidizing agents.
(B) They form ions with a +1 oxidation state.
(C) As the atomic numbers of the alkali metals increase, the electronegativity decreases.
(D) As the atomic numbers of the alkali metals increase, their first ionization energy decreases.

38. As a beaker of water is heated over a flame, the temperature increases steadily until it reaches 373 K. At that point, the beaker is left on the open flame, but the temperature remains at 373 K as long as water remains in the beaker. This is because at 373 K, the energy supplied by the flame

(A) is completely absorbed by the glass beaker.
(B) is less than the energy lost by the water through electromagnetic radiation.
(C) is dissipated by the water as visible light.
(D) is used to overcome the heat of vaporization of the water.

39. For which of the following reactions will the equilibrium constants K_c and K_p have the same value?

(A) $2\,N_2O_5(g) \leftrightarrow 2\,NO_2(g) + O_2(g)$
(B) $2\,CO_2(g) \leftrightarrow 2\,CO(g) + O_2(g)$
(C) $H_2O\,(g) + CO(g) \leftrightarrow H_2(g) + CO_2(g)$
(D) $3\,O_2(g) \leftrightarrow 2\,O_3(g)$

40. Each of the following compounds was added to distilled water at 25°C. Which one produced a solution with a pH that was less than 7?

(A) O_2
(B) NaI
(C) MgO
(D) SO_2

GO ON TO THE NEXT PAGE.

41. ...Be$_2$C + ...H$_2$O → ...Be(OH)$_2$ + ...CH$_4$

    When the equation for the reaction above is balanced with the lowest whole-number coefficients, the coefficient for H$_2$O will be

    (A) 1
    (B) 2
    (C) 3
    (D) 4

42. A sample of an ideal gas is placed in a sealed container of constant volume. If the temperature of the gas is increased from 40°C to 70°C, which of the following values for the gas will NOT increase?

    (A) The average speed of the molecules of the gas
    (B) The average kinetic energy of the molecules of the gas
    (C) The pressure exerted by the gas
    (D) The density of the gas

43. A $^{222}_{86}$Rn nuclide decays through the emission of two beta particles and two alpha particles. The resulting nuclide is

    (A) $^{214}_{84}$Po
    (B) $^{210}_{84}$Po
    (C) $^{214}_{83}$Bi
    (D) $^{210}_{83}$Bi

44. When 80.0 ml of a 0.40 M NaI solution is combined with 20.0 ml of a 0.30 M CaI$_2$ solution, what will be the molar concentration of I$^-$ ions in the solution?

    (A) 0.70 M
    (B) 0.44 M
    (C) 0.38 M
    (D) 0.35 M

45. Hydrogen sulfide is a toxic waste product of some industrial processes. It can be recognized by its distinctive rotten egg smell. One method for elimininating hydrogen sulfide is by reaction with dissolved oxygen, as shown below.

    $$2\,H_2S + O_2 \rightarrow 2\,S + 2\,H_2O$$

    If 102 grams of H$_2$S are combined with 64 grams of O$_2$, what is the maximum mass of elemental sulfur that could be produced by the reaction?

    (A) 32 grams
    (B) 48 grams
    (C) 64 grams
    (D) 96 grams

46.
$$H_2O(g) \rightarrow H_2O(l)$$

    Which of the following is true of the values of ΔH, ΔS, and ΔG for the reaction shown above at 25°C?

	ΔH	ΔS	ΔG
(A)	Positive	Positive	Positive
(B)	Positive	Negative	Negative
(C)	Negative	Negative	Positive
(D)	Negative	Negative	Negative

47. A student added 1 liter of a 1.0 M Na$_2$SO$_4$ solution to 1 liter of a 1.0 M Ag(C$_2$H$_3$O$_2$) solution. A silver sulfate precipitate formed, and nearly all of the silver ions disappeared from the solution. Which of the following lists the ions remaining in the solution in order of decreasing concentration?

    (A) [SO$_4^{2-}$] > [C$_2$H$_3$O$_2^-$] > [Na$^+$]
    (B) [C$_2$H$_3$O$_2^-$] > [Na$^+$] > [SO$_4^{2-}$]
    (C) [Na$^+$] > [SO$_4^{2-}$] > [C$_2$H$_3$O$_2^-$]
    (D) [Na$^+$] > [C$_2$H$_3$O$_2^-$] > [SO$_4^{2-}$]

48. Iron ore is converted to pure iron by the following reaction, which occurs at extremely high temperatures:

    $$Fe_2O_3 + 3\,CO \rightarrow 2\,Fe + 3\,CO_2$$

    A 1,600 gram sample of iron ore reacted completely to form 558 grams of pure iron. What was the percent of Fe$_2$O$_3$ (molecular weight 160) by mass in the original sample?

    (A) 25%
    (B) 33%
    (C) 50%
    (D) 100%

49. Gold and silver have been used throughout history to make coins. Which of the following statements could account for the popularity of these metals for use in coins?

    (A) Network bonds make these metals especially durable.
    (B) Negative oxidation potentials for these metals make them especially unreactive with their surroundings.
    (C) These metals are among the lightest of the elements.
    (D) These metals are commonly found in nature.

GO ON TO THE NEXT PAGE.

50.
$$2\ HI(g) \rightarrow H_2(g) + I_2(g)$$

For the reaction given above, ΔH^o is –50 kJ. Based on the information given in the table below, what is the average bond energy of the H–I bond?

Bond	Average Bond Energy (kJ/mol)
H–H	440
I–I	150

(A) 270 kJ/mol
(B) 540 kJ/mol
(C) 590 kJ/mol
(D) 640 kJ/mol

51. All of the elements listed below are gases at room temperature. Which gas would be expected to show the greatest deviation from ideal behavior?

(A) He
(B) Ar
(C) Kr
(D) Xe

Questions 52 and 53 refer to the information given below.

$$F_2(g) + 2\ ClO_2(g) \rightarrow 2\ FClO_2(g)$$

Experiment	$[F_2]$ (M)	$[ClO_2]$ (M)	Initial rate of disappearance of F_2 (M/ sec)
1	0.10	0.010	1.2×10^{-3}
2	0.20	0.010	2.4×10^{-3}
3	0.40	0.020	9.6×10^{-3}

52. Based on the data given in the table, which of the following expressions is equal to the rate law for the reaction given above?

(A) Rate = $k[F_2]$
(B) Rate = $k[ClO_2]$
(C) Rate = $k[F_2][ClO_2]$
(D) Rate = $k[F_2]^2[ClO_2]$

53. What is the initial rate of disappearance of ClO_2 in experiment 2?

(A) 1.2×10^{-3} M/sec
(B) 2.4×10^{-3} M/sec
(C) 4.8×10^{-3} M/sec
(D) 7.0×10^{-3} M/sec

54. Hypobromous acid, HBrO, is added to distilled water. If the acid dissociation constant for HBrO is equal to 2×10^{-9}, what is the concentration of HBrO when the pH of the solution is equal to 5?

(A) 5-molar
(B) 1-molar
(C) 0.1-molar
(D) 0.05-molar

55. At 25°C, the vapor pressure of water is 24 mmHg. Which of the following expressions gives the vapor pressure of a solution created by adding 2.0 moles of sucrose to 55 moles of water?

(A) $\dfrac{(24)(2.0)}{(55)}$ mmHg

(B) $\dfrac{(2.0)}{(24)(55)}$ mmHg

(C) $\dfrac{(24)(55)}{(57)}$ mmHg

(D) $\dfrac{(55)}{(24)(57)}$ mmHg

56.
$$PCl_5(g) \longleftrightarrow PCl_3(g) + Cl_2(g)$$

When PCl_5 is placed in an evacuated container at 250°C, the above reaction takes place. The pressure in the container before the reaction takes place is 6 atm and is entirely due to the PCl_5 gas. After the reaction comes to equilibrium, the partial pressure due to Cl_2 is found to be 2 atm. What is the value of the equilibrium constant, K_p, for this reaction?

(A) 0.5
(B) 1
(C) 2
(D) 5

57.
$Co^{2+} + 2\ e^- \rightarrow Co$	$E^o = -0.28$ V
$Cr^{3+} + e^- \rightarrow Cr^{2+}$	$E^o = -0.41$ V

According to the information shown above, which of the following statements is true regarding the reaction below?

$$2\ Co + Cr^{3+} \rightarrow 2\ Co^{2+} + Cr^{2+}$$

(A) Cr^{3+} acts as the reducing agent.
(B) The reaction would take place in a galvanic cell.
(C) K_{eq} for the reaction is less than 1.
(D) Cobalt metal is reduced in the reaction.

GO ON TO THE NEXT PAGE.

58. What is the pH of a solution made by mixing 200 milliliters of a 0.20-molar solution of NH_3 with 200 milliliters of a 0.20-molar of an NH_4Cl solution? (The base dissociation constant, K_b, for NH_3 is 1.8×10^{-5}.)

   (A) Between 4 and 5
   (B) Between 5 and 6
   (C) Between 8 and 9
   (D) Between 9 and 10

59. $$POCl_3(l) \rightarrow POCl_3(g)$$

   For the reaction above, ΔH is 50 kilojoules per mole and ΔS is 100 joules per mole. What is the boiling point of $POCl_3$? (Assume that ΔH and ΔS remain constant with changing temperature.)

   (A)   2 K
   (B)   50 K
   (C) 200 K
   (D) 500 K

60. When fluorine gas is bubbled through a concentrated aqueous solution of potassium chloride at room temperature, which of the following would be expected to occur?

   (A) A reaction will occur, and chlorine gas will be produced.
   (B) A reaction will occur, and potassium fluoride will precipitate.
   (C) A reaction will occur, and solid chlorine will precipitate.
   (D) A reaction will occur, and solid potassium will precipitate.

END OF SECTION 1

CHEMISTRY
SECTION II
Time—1 hour and 30 minutes

<u>General Instructions</u>

Calculators, including those with programming and graphing capabilities, may be used in Part A. However, calculators with typewriter-style (QWERTY) keyboards are NOT permitted.

Pages containing a periodic table, the electrochemical series, and equations commonly used in chemistry will be available for your use.

You may write your answers with either a pen or a pencil. Be sure to write CLEARLY and LEGIBLY. If you make an error, you may save time by crossing it out rather than trying to erase it.

Write all your answers in the essay booklet. Number your answers as the questions are numbered in the examination booklet.

GO ON TO THE NEXT PAGE.

MATERIAL IN THE FOLLOWING TABLE AND IN THE TABLES ON THE NEXT 3 PAGES MAY BE USEFUL IN
ANSWERING THE QUESTIONS IN THIS SECTION OF THE EXAMINATION.

PERIODIC CHART OF THE ELEMENTS

1 **H** 1.0																	2 **He** 4.0
3 **Li** 6.9	4 **Be** 9.0											5 **B** 10.8	6 **C** 12.0	7 **N** 14.0	8 **O** 16.0	9 **F** 19.0	10 **Ne** 20.2
11 **Na** 23.0	12 **Mg** 24.3											13 **Al** 27.0	14 **Si** 28.1	15 **P** 31.0	16 **S** 32.1	17 **Cl** 35.5	18 **Ar** 39.9
19 **K** 39.1	20 **Ca** 40.1	21 **Sc** 45.0	22 **Ti** 47.9	23 **V** 50.9	24 **Cr** 52.0	25 **Mn** 54.9	26 **Fe** 55.8	27 **Co** 58.9	28 **Ni** 58.7	29 **Cu** 63.5	30 **Zn** 65.4	31 **Ga** 69.7	32 **Ge** 72.6	33 **As** 74.9	34 **Se** 79.0	35 **Br** 79.9	36 **Kr** 83.8
37 **Rb** 85.5	38 **Sr** 87.6	39 **Y** 88.9	40 **Zr** 91.2	41 **Nb** 92.9	42 **Mo** 95.9	43 **Tc** (98)	44 **Ru** 101.1	45 **Rh** 102.9	46 **Pd** 106.4	47 **Ag** 107.9	48 **Cd** 112.4	49 **In** 114.8	50 **Sn** 118.7	51 **Sb** 121.8	52 **Te** 127.6	53 **I** 126.9	54 **Xe** 131.3
55 **Cs** 132.9	56 **Ba** 137.3	57 ***La** 138.9	72 **Hf** 178.5	73 **Ta** 180.9	74 **W** 183.9	75 **Re** 186.2	76 **Os** 190.2	77 **Ir** 192.2	78 **Pt** 195.1	79 **Au** 197.0	80 **Hg** 200.6	81 **Tl** 204.4	82 **Pb** 207.2	83 **Bi** 209.0	84 **Po** (209)	85 **At** (210)	86 **Rn** (222)
87 **Fr** (223)	88 **Ra** 226.0	89 **†Ac** 227.0															

*Lanthanum Series

58 **Ce** 140.1	59 **Pr** 140.9	60 **Nd** 144.2	61 **Pm** (145)	62 **Sm** 150.4	63 **Eu** 152.0	64 **Gd** 157.3	65 **Tb** 158.9	66 **Dy** 162.5	67 **Ho** 164.9	68 **Er** 167.3	69 **Tm** 168.9	70 **Yb** 173.0	71 **Lu** 175.0

†Actinium Series

90 **Th** 232.0	91 **Pa** 231.0	92 **U** 238.0	93 **Np** 237.0	94 **Pu** (244)	95 **Am** (243)	96 **Cm** (247)	97 **Bk** (247)	98 **Cf** (251)	99 **Es** (252)	100 **Fm** (258)	101 **Md** (258)	102 **No** (259)	103 **Lr** (260)

DO NOT DETACH FROM BOOK.

GO ON TO THE NEXT PAGE.

ADVANCED PLACEMENT CHEMISTRY EQUATIONS AND CONSTANTS

ATOMIC STRUCTURE

$$E = h\nu \qquad\qquad c = \lambda\nu$$

$$\lambda = \frac{h}{mv} \qquad\qquad p = mv$$

$$E_n = \frac{-2.178 \times 10^{-18}}{n^2} \text{ joule}$$

EQUILIBRIUM

$$K_a = \frac{[H^-][A^-]}{[HA]}$$

$$K_b = \frac{[OH^-][HB^+]}{[B]}$$

$$K_w = [OH^-][H^+] = 10^{-14} @ 25°C$$

$$= K_a \times K_b$$

$$pH = -\log [H^+], pOH = -\log [OH^-]$$

$$14 = pH + pOH$$

$$pH = pK_a + \log \frac{[A^-]}{[HA]}$$

$$pOH = pK_b + \log \frac{[HB^+]}{[B]}$$

$$pK_a = -\log K_a, pK_b = -\log K_b$$

$$K_p = K_c(RT)^{\Delta n},$$

where Δn = moles product gas – moles reactant gas

THERMOCHEMISTRY / KINETICS

$$\Delta S° = \sum S°_{products} - \sum S°_{reactants}$$

$$\Delta H° = \sum \Delta H°_f{}_{products} - \sum \Delta H°_f{}_{reactants}$$

$$\Delta G° = \sum \Delta G°_f{}_{products} - \sum \Delta G°_f{}_{reactants}$$

$$\Delta G° = \Delta H° - T\Delta S°$$

$$= -RT \ln K = -2.303 \, RT \log K$$

$$= -nFE°$$

$$\Delta G = \Delta G° + RT \ln Q = \Delta G° + 2.303 \, RT \log Q$$

$$q = mc\Delta T$$

$$C_p = \frac{\Delta H}{\Delta T}$$

$$\ln[A]_t - \ln[A]_o = -kt$$

$$\frac{1}{[A]_t} - \frac{1}{[A]_o} = kt$$

$$\ln k = \frac{-E_a}{R}\left(\frac{1}{T}\right) + \ln A$$

E = energy $\qquad v$ = velocity

ν = frequency $\qquad n$ = principal quantum number

λ = wavelength $\qquad m$ = mass

p = momentum

Speed of light, $c = 3.00 \times 10^8$ m s^{-1}

Planck's constant, $h = 6.63 \times 10^{-34}$ J s

Boltzmann's constant, $k = 1.38 \times 10^{-23}$ J K^{-1}

Avogadro's number $= 6.022 \times 10^{23}$ mol^{-1}

Electron charge, $e = -1.602 \times 10^{-19}$ coulomb

1 electron volt/atom $= 96.5$ kJ mol^{-1}

Equilibrium Constants

K_a (weak acid)
K_b (weak base)
K_w (water)
K_p (gas pressure)
K_c (molar concentrations)

$S°$ = standard entropy
$H°$ = standard enthalpy
$G°$ = standard free energy
$E°$ = standard reduction potential
T = temperature
n = moles
m = mass
q = heat
c = specific heat capacity
C_p = molar heat capacity at constant pressure
E_a = activation energy
k = rate constant
A = frequency factor

Faraday's Constant, F = 96,500 coulombs per mole
of electrons

Gas Constant, R = 8.31 J mol^{-1}K^{-1}
= 0.0821 L atm mol^{-1}K^{-1}
= 8.31 volt coulomb mol^{-1}K^{-1}

GO ON TO THE NEXT PAGE.

ADVANCED PLACEMENT CHEMISTRY EQUATIONS AND CONSTANTS

GASES, LIQUIDS, AND SOLUTIONS

$$PV = nRT$$

$$\left(P + \frac{n^2 a}{V^2}\right)(V - nb) = nRT$$

$$P_A = P_{total} \cdot X_A, \text{ where } X_A = \frac{\text{moles A}}{\text{total moles}}$$

$$P_{total} = P_A + P_B + P_C{}^+ ...$$

$$n = \frac{m}{M}$$

$$K = {}^\circ C + 273$$

$$\frac{P_1 V_1}{T_1} = \frac{P_2 V_2}{T_2}$$

$$D = \frac{m}{V}$$

$$U_{rms} = \sqrt{\frac{3kT}{m}} = \sqrt{\frac{3RT}{M}}$$

$$KE \text{ per molecule} = \frac{1}{2}mv^2$$

$$KE \text{ per mole} = \frac{3}{2}RT$$

$$\frac{r_1}{r_2} = \sqrt{\frac{M_2}{M_1}}$$

molarity, M = moles solute per liter solution
molality = moles solute per kilogram solvent

$$\Delta T_f = iK_f \times \text{molality}$$

$$\Delta T_b = iK_b \times \text{molality}$$

$$\pi = MRT$$

$$A = abc$$

OXIDATION REDUCTION; ELECTROCHEMISTRY

$$Q = \frac{[C]^c\,[D]^d}{[A]^a\,[B]^b} \text{ where } a\,A + b\,B \rightarrow c\,C + d\,D$$

$$I = \frac{q}{t}$$

$$E_{cell} = E^\circ{}_{cell} - \frac{RT}{nF}\ln Q = E^\circ{}_{cell} - \frac{0.0592}{n}\log Q \text{ @ } 25^\circ C$$

$$\log K = \frac{nE^\circ}{0.0592}$$

P = pressure
V = volume
T = temperature
n = number of moles
D = density
m = mass
v = velocity

U_{rms} = root-mean-square speed
KE = kinetic energy
r = rate of effusion
M = molar mass
π = osmotic pressure
i = van't Hoff factor
K_f = molal freezing-point depression constant
K_b = molal boiling-point elevation constant
A = absorbance
a = molar absorptivity
b = path length
c = concentration
Q = reaction quotient
l = current (amperes)
q = charge (coulombs)
t = time (seconds)
E° = standard reduction potential
K = equilibrium constant

$$\begin{aligned}
\text{Gas constant, } R &= 8.31 \text{ J mol}^{-1}\text{K}^{-1} \\
&= 0.0821 \text{ L atm mol}^{-1}\text{ K}^{-1} \\
&= 8.31 \text{ volt coulomb mol}^{-1}\text{K}^{-1}
\end{aligned}$$

Boltzmann's constant, $k = 1.38 \times 10^{-23}$ J K^{-1}
K_f for H_2O = 1.86 K kg mol^{-1}
K_b for H_2O = 0.512 K kg mol^{-1}
1 atm = 760 mm Hg
= 760 torr
STP = 0.000°C and 1.000 atm
1 faraday, F = 96,500 coulumbs per mole of electrons

GO ON TO THE NEXT PAGE.

STANDARD REDUCTION POTENTIALS IN AQUEOUS SOLUTION AT 25 C (in V)			
$F_2(g) + 2\,e^-$	$\rightarrow$	$2\,F^-$	2.87
$Co^{3+} + e^-$	$\rightarrow$	Co^{2+}	1.82
$Au^{3+} + 3e^-$	$\rightarrow$	$Au(s)$	1.50
$Cl_2(g) + 2\,e^-$	$\rightarrow$	$2\,Cl^-$	1.36
$O_2(g) + 4H^+ + 4\,e^-$	$\rightarrow$	$2\,H_2O$	1.23
$Br_2(l) + 2\,e^-$	$\rightarrow$	$2\,Br^-$	1.07
$2\,Hg^{2+} + 2\,e^-$	$\rightarrow$	Hg_2^{2+}	0.92
$Hg^{2+} + 2\,e^-$	$\rightarrow$	$Hg(l)$	0.85
$Ag^+ + e^-$	$\rightarrow$	$Ag(s)$	0.80
$Hg_2^{2+} + 2\,e^-$	$\rightarrow$	$2\,Hg(l)$	0.79
$Fe^{3+} + e^-$	$\rightarrow$	Fe^{2+}	0.77
$I_2(s) + 2\,e^-$	$\rightarrow$	$2\,I^-$	0.53
$Cu^+ + e^-$	$\rightarrow$	$Cu(s)$	0.52
$Cu^{2+} + 2\,e^-$	$\rightarrow$	$Cu(s)$	0.34
$Cu^{2+} + e^-$	$\rightarrow$	Cu^+	0.15
$Sn^{4+} + 2\,e^-$	$\rightarrow$	Sn^{2+}	0.15
$S(s) + 2\,H^+ + 2\,e^-$	$\rightarrow$	H_2S	0.14
$2\,H^+ + 2\,e^-$	$\rightarrow$	$H_2(g)$	0.00
$Pb^{2+} + 2\,e^-$	$\rightarrow$	$Pb(s)$	−0.13
$Sn^{2+} + 2\,e^-$	$\rightarrow$	$Sn(s)$	−0.14
$Ni^{2+} + 2\,e^-$	$\rightarrow$	$Ni(s)$	−0.25
$Co^{2+} + 2\,e^-$	$\rightarrow$	$Co(s)$	−0.28
$Tl^+ + e^-$	$\rightarrow$	$Tl(s)$	−0.34
$Cd^{2+} + 2\,e^-$	$\rightarrow$	$Cd(s)$	−0.40
$Cr^{3+} + e^-$	$\rightarrow$	Cr^{2+}	−0.41
$Fe^2 + 2\,e^-$	$\rightarrow$	$Fe(s)$	−0.44
$Cr^{3+} + 3\,e^-$	$\rightarrow$	$Cr(s)$	−0.74
$Zn^{2+} + 2\,e^-$	$\rightarrow$	$Zn(s)$	−0.76
$Mn^{2+} + 2\,e^-$	$\rightarrow$	$Mn(s)$	−1.18
$Al^{3+} + 3\,e^-$	$\rightarrow$	$Al(s)$	−1.66
$Be^{2+} + 2\,e^-$	$\rightarrow$	$Be(s)$	−1.70
$Mg^{2+} + 2\,e^-$	$\rightarrow$	$Mg(s)$	−2.37
$Na^+ + e^-$	$\rightarrow$	$Na(s)$	−2.71
$Ca^{2+} + 2\,e^-$	$\rightarrow$	$Ca(s)$	−2.87
$Sr^{2+} + 2\,e^-$	$\rightarrow$	$Sr(s)$	−2.89
$Ba^{2+} + 2\,e^-$	$\rightarrow$	$Ba(s)$	−2.90
$Rb^+ + e^-$	$\rightarrow$	$Rb(s)$	−2.92
$K^+ + e^-$	$\rightarrow$	$K(s)$	−2.92
$Cs^+ + e^-$	$\rightarrow$	$Cs(s)$	−2.92
$Li^+ + e^-$	$\rightarrow$	$Li(s)$	−3.05

GO ON TO THE NEXT PAGE.

CHEMISTRY

Section II

(Total time—90 minutes)

Part A

YOU MAY USE YOUR CALCULATOR FOR PART A.

THE METHODS USED AND THE STEPS INVOLVED IN ARRIVING AT YOUR ANSWERS MUST BE SHOWN CLEARLY. It is to your advantage to do this since you may obtain partial credit if you do, and you will receive little or no credit if you do not. Attention should be paid to significant figures.

Be sure to write your answers in the space provided following each question.

Answer Questions 1, 2, and 3.

1. At 25°C, the solubility product consant, K_{sp}, for nickel hydroxide, $Ni(OH)_2$, is 1.6×10^{-14}.

    (a) Write a balanced equation for the solubility equilibrium for $Ni(OH)_2$.

    (b) What is the molar solubility of $Ni(OH)_2$ in pure water at 25°C?

    (c) A 1.0-molar NaOH solution is slowly added to a saturated solution of $Ni(OH)_2$ at 25°C. If excess solid $Ni(OH)_2$ remains in the solution throughout the procedure, what is the concentration of Ni^{2+} ions in the solution at the moment when the pH is equal to 11?

    (d) Predict whether a precipitate will form when 200.0 milliliters of a 5.0×10^{-5}-molar KOH solution is mixed with 300.0 milliliters of a 2.0×10^{-4}-molar $Ni(NO_3)_2$ solution at 25°C. Show calculations to support your prediction.

    (e) At 25°C, 100 ml of a saturated $Ni(OH)_2$ solution was prepared.

    (i) Calculate the mass of $Ni(OH)_2$ present in the solution.

    (ii) Calculate the pH of the solution.

    (iii) If the solution is allowed to evaporate to a final volume of 50 ml, what will be the pH? Justify your answer.

GO ON TO THE NEXT PAGE.

2. In two separate experiments, a sample of an unknown hydrocarbon was burned in air, and a sample of the same hydrocarbon was placed into an organic solvent.

   (a) When the hydrocarbon sample was burned in a reaction that went to completion, 2.2 grams of water and 3.6 liters of carbon dioxide were produced under standard conditions. What is the empirical formula of the hydrocarbon?

   (b) When 4.05 grams of the unknown hydrocarbon was placed in 100.0 grams of benzene, C_6H_6, the freezing point of the solution was measured to be 1.66°C. The normal freezing point of benzene is 5.50°C and the freezing-point depression constant for benzene is 5.12°C/m. What is the molecular weight of the unknown hydrocarbon?

   (c) What is the molecular formula and name of the hydrocarbon?

   (d) Write the balanced equation for the combustion reaction that took place in (a).

   (e) Draw two isomers for the hydrocarbon.

3. A strip of Ni metal is placed in a 1-molar solution of $Ni(NO_3)_2$ and a strip of Ag metal is placed in a 1-molar solution of $AgNO_3$. An electrochemical cell is created when the two solutions are connected by a salt bridge and the two metal strips are connected by wires to a voltmeter.

   (a) Write the balanced chemical equation for the overall reaction that occurs in the cell, and calculate the cell potential, $E°$.

   (b) Calculate how many grams of metal will be deposited on the cathode if the cell is allowed to run at a constant current of 1.5 amperes for 8.00 minutes.

   (c) Calculate the value of the standard free energy change, $\Delta G°$, for the cell reaction.

   (d) Calculate the cell potential, E, at 25°C for the cell shown above if the initial concentration of $Ni(NO_3)_2$ is 0.100-molar and the initial concentration of $AgNO_3$ is 1.20-molar.

   (e) Is the reaction in the cell spontaneous under conditions described in part (d)? Justify your answer.

GO ON TO THE NEXT PAGE.

Part B

Time—40 minutes

CALCULATORS MAY BE USED FOR PART B

4. You will be given three chemical reactions below. In part (i), write the balanced equation for the reaction, leaving coefficients in terms of lowest whole numbers. Then answer the question pertaining to that reaction in part (ii). For each of the following three reactions, assume that solutions are aqueous unless it says otherwise. Substances in solutions should be represented as ions if these substances are extensively ionized. Omit formulas for ions or molecules that are not affected by the reaction. Only equations inside the answer boxes will be graded.

EXAMPLE:

A piece of solid zinc is placed in a solution of silver(I) acetate

> (i) Write the balanced equation in this box:
>
> $$Zn + 2Ag^+ \rightarrow Zn^{2+} + 2Ag$$

> (ii) Which substance is reduced in the reaction?
>
> Ag$^+$ is reduced

(a) Solutions of sodium acetate and nitric acid are mixed.

> (i) Write the balanced equation in this box:

> (ii) Identify the spectator ion in this reaction.

GO ON TO THE NEXT PAGE.

(b)   A piece of solid calcium is heated in oxygen gas.

> (i) Write the balanced equation in this box:

> (ii) Is the entropy change for this reaction positive or negative? Briefly explain.

_____

_____

(c)   Ammonia and boron trichloride gases are mixed.

> (i) Write the balanced equation in this box:

> (ii) Identify the Lewis acid in this reaction.

_____

_____

GO ON TO THE NEXT PAGE.

Answer Question 5 and Question 6.

Answering these questions provides an opportunity to demonstrate your ability to present your material in logical, coherent, and convincing English. Your responses will be judged on the basis of accuracy and importance of the detail cited and on the appropriateness of the descriptive material used. Specific answers are preferable to broad, diffuse responses. Illustrative examples and equations may be helpful.

5. Use your knowledge of chemical principles to answer the following questions.

	First ionization energy (kJ/mol)	Second ionization energy (kJ/mol)
Na	490	4,560
K	420	3,050

(a) The table above shows the first and second ionization energies for potassium and sodium.

  (i) Explain the difference between the first ionization energies of potassium and sodium.

  (ii) Explain why the second ionization energy for potassium is so much larger than the first ionization energy.

	Atomic radius (pm)
Na	190
Na^+	120
K	230
K^+	150
Cl	100
Cl^-	170
Br	110
Br^-	190

(b) The table above shows the atomic radii of several atoms and ions.

  (i) The atomic radius of bromine is smaller than the atomic radius of potassium. Explain.

  (ii) The potassium and chlorine ions have the same electronic structure. Explain the difference in their atomic radii.

  (iii) When potassium bromide and sodium chloride are compared, which will have the higher melting point? Explain.

GO ON TO THE NEXT PAGE.

Name	Formula	Boiling point (°C)
Methane	CH_4	−164
Methoxymethane	CH_3OCH_3	−23
Ethanol	CH_3CH_2OH	78

(c) The table above shows the boiling points for several organic compounds. Explain the trend in increasing boiling point.

6. An experiment is to be performed involving the titration of a solution of benzoic acid, $HC_7H_5O_2$, of unknown concentration with a standardized 1.00-molar NaOH solution. The pK_a of benzoic acid is 4.2.

(a) What measurements and calculations must be made if the concentration of the acid is to be determined?

(b) An indicator must be chosen for the titration. If methyl red (pH interval for color change: 4.2–6.3) and thymol blue (pH interval for color change: 8.0–9.6) are the only indicators available, which one should be used? Explain your choice.

(c) Describe how you would use the titration apparatus and benzoic acid and NaOH solutions to create a buffer solution with a pH of 4.2 after you had measured the volume of NaOH solution required to reach the equivalence point for a given volume of benzoic acid solution.

(d) Describe a way to test whether the neutralization of benzoic acid by a strong base is an exothermic process, using readily available school laboratory equipment.

GO ON TO THE NEXT PAGE.

Answer Question 7.

7.

$$H_3PO_4 \rightleftharpoons H^+ + H_2PO_4^- \qquad K_1 = 7.5 \times 10^{-3}$$

$$H_2PO_4^- \rightleftharpoons H^+ + HPO_4^{2-} \qquad K_2 = 6.2 \times 10^{-8}$$

$$HPO_4^{2-} \rightleftharpoons H^+ + PO_4^{3-} \qquad K_3 = 2.2 \times 10^{-13}$$

(a) Choose an amphoteric species from the reactions listed above, and give its conjugate acid and its conjugate base.

(b) Explain why the dissociation constant decreases with each hydrogen ion lost.

(c) Of the acids listed above, which would be most useful in creating a buffer solution with a pH of 7.5?

(d) Sketch the titration curve that results when H_3PO_4 is titrated with excess NaOH and label the two axes.

STOP

END OF EXAM

Chapter 13
Practice Test 2
Answers and
Explanations

ANSWER KEY EXAM 2

1.	D	31.	A
2.	A	32.	B
3.	D	33.	A
4.	C	34.	C
5.	D	35.	D
6.	B	36.	D
7.	C	37.	A
8.	D	38.	D
9.	A	39.	C
10.	C	40.	D
11.	D	41.	D
12.	A	42.	D
13.	D	43.	A
14.	B	44.	B
15.	C	45.	D
16.	D	46.	D
17.	A	47.	D
18.	A	48.	C
19.	C	49.	B
20.	B	50.	A
21.	D	51.	D
22.	D	52.	C
23.	D	53.	C
24.	A	54.	D
25.	B	55.	C
26.	C	56.	B
27.	B	57.	C
28.	C	58.	D
29.	B	59.	D
30.	B	60.	A

QUESTIONS	EXPLANATIONS

SECTION I—MULTIPLE CHOICE

Questions 1–3 refer to the following elements.

    (A) Ne
    (B) Li
    (C) Al
    (D) Cl

1. Which of the elements above is most commonly found as a negatively charged ion?

1. **D** Chlorine (Cl) forms a negatively charged ion with an oxidation state of –1. None of the other elements listed normally forms negative ions.

2. Which of the elements above has the same electron structure as Mg^{2+}?

2. **A** Mg^{2+} has the following electron structure: $1s^2\ 2s^2 2p^6$. That's the electron structure that neutral neon (Ne) has.

3. Which of the elements above is found as a diatomic gas in its uncombined state?

3. **D** Chlorine is found as a diatomic gas, Cl_2, in its uncombined state.

Questions 4–7 refer to the following substances.

    (A) H_2O
    (B) NH_3
    (C) CH_4
    (D) HF

4. This molecule is nonpolar.

4. **C** CH_4 has a symmetrical tetrahedral shape and is nonpolar.

All of the other choices are polar, as shown on the following page.

(A)

(−)

H   H

(+)

(B)

(−)

H   H

(+)

(E)

(−)

H   H

(+)

(D)   (+) H–F (−)

5. This substance forms an aqueous solution with a pH that's less than 7.

5. **D** HF is the only acid on the list.

6. This molecule has a trigonal pyramidal shape.

6. **B** NH_3 is the only choice with a trigonal pyramidal shape, and it has one lone electron pair. For the shapes of the other molecules, see the answer to Question 4.

QUESTIONS	EXPLANATIONS

QUESTIONS

7. This substance does NOT exhibit hydrogen bonding.

Questions 8 and 9 refer to an experiment in which five individual 1-liter aqueous solutions, each containing a 1-mole sample of one of the salts listed below, were subjected to various tests at room temperature.

(A) $NaC_2H_3O_2$
(B) $NaCl$
(C) $MgBr_2$
(D) $HC_2H_3O_2$

8. The solution containing this salt had the lowest conductivity.

9. The solution containing this salt had the highest pH.

10. Titanium metal is prepared by heating rutile, an oxide of titanium, along with carbon and chlorine gas. By mass, rutile is 40% oxygen and 60% titanium. What is the empirical formula of rutile?

(A) TiO
(B) Ti_2O
(C) TiO_2
(D) Ti_2O_3

EXPLANATIONS

7. **C** Remember FON. Only hydrogen compounds that contain fluorine, oxygen, or nitrogen exhibit hydrogen bonding.

8. **D** The conductivity of a solution depends on the number of charged particles it contains. $HC_2H_3O_2$ is a weak acid, so very little of the one-molar sample will dissociate in solution, producing very few charged particles. All of the other choices are soluble salts, which dissociate completely.

9. **A** $NaC_2H_3O_2$ is composed of the conjugates of a strong base ($NaOH$) and a weak acid ($HC_2H_3O_2$). It will produce a basic solution that will have a pH greater than 7. Choices (B) and (C) will produce neutral solutions with pHs of 7, and choice (D) will produce an acidic solution, with a pH of less than 7.

10. **C** Let's say you have 100 grams of the rutile. So you have 40 grams of oxygen and 60 grams of titanium. Now convert to moles.

$$\text{Moles} = \frac{\text{grams}}{\text{MW}}$$

$$\text{Moles of oxygen} = \frac{(40\text{ g})}{(16\text{ g/mol})} = 2.5$$

$$\text{Moles of titanium} = \frac{(60\text{ g})}{(48\text{ g/mol})} = 1.25$$

The molar ratio of O to Ti is 2 to 1, so the empirical formula must be TiO_2.

QUESTIONS	EXPLANATIONS

11. What is the weight of NaNO₃ (molecular weight 85.0) present in 100.0 ml of a 4.00-molar solution?

    (A) 8.50 grams
    (B) 17.0 grams
    (C) 25.5 grams
    (D) 34.0 grams

11. D Moles = (molarity)(volume)

Moles of NaNO₃ = (4.00 M)(0.100 L)
             = 0.400 moles

Grams = (moles)(MW)

Grams of NaNO₃ = (0.400 moles)(85.0 g/mol)
             = 34.0 grams

12. A mixture of gases contains 5.0 moles of oxygen, 12 moles of nitrogen, and 4.0 moles of carbon dioxide. If the partial pressure due to carbon dioxide is 1.6 atmospheres, what is the partial pressure due to oxygen?

    (A) 2.0 atm
    (B) 4.8 atm
    (C) 6.0 atm
    (D) 6.8 atm

12. A From Dalton's law, the partial pressure of a gas in a sample is directly proportional to its molar quantity. So if 4 moles of carbon dioxide have a partial pressure of 1.6 atm, then 5 moles of oxygen will have a partial pressure of 2.0 atm.

13. Which of the diagrams below represents the enthalpy change for an endothermic reaction?

13. D The change in enthalpy, Δ*H*, is the difference in potential energy between the final and initial states of a reaction.

(A)

(C)

(B)

(D)

14.   2 NO(*g*) + Br₂(*g*) → 2 NOBr(*g*)

For the reaction above, the experimental rate law is given as follows:

$$Rate = k[NO]^2[Br_2]$$

Which of the statements below is true regarding this reaction?

    (A) The reaction is first-order overall.
    (B) The reaction is first-order with respect to Br₂.
    (C) The reaction is first-order with respect to NO.
    (D) The reaction is second-order overall.

14. B The exponent for Br₂ in the rate law is equal to 1, so the reaction is first-order with respect to Br₂. By the way, the reaction is second-order with respect to NO and third-order overall.

QUESTIONS	EXPLANATIONS

15. Which of the following is the most likely electron configuration of a sulfur atom in its ground state?

(A) $1s^2\,2s^22p^6$
(B) $1s^2\,2s^22p^63s^23p^2$
(C) $1s^2\,2s^22p^63s^23p^4$
(D) $1s^2\,2s^22p^63s^23p^6$

15. **C** $1s^2\,2s^22p^6\,3s^23p^4$ is the proper ground-state configuration for sulfur's 16 electrons.

16. How many grams of carbon are present in 270 grams of glucose, $C_6H_{12}O_6$?

(A) 12.0 grams
(B) 18.0 grams
(C) 67.5 grams
(D) 72.0 grams

16. **D** $\text{Moles} = \dfrac{\text{grams}}{\text{MW}}$

$\text{Moles of glucose} = \dfrac{(270\text{ g})}{(180\text{ g/mol})} = 1.5\text{ moles}$

1.50 moles of $C_6H_{12}O_6$ contains 9.00 moles of carbon.

$\text{Grams} = (\text{moles})(\text{MW})$

$\text{Grams of carbon} = (9.00\text{ moles})(12.0\text{ grams/mol})$
$\qquad\qquad\qquad = 108\text{ grams}$

You could also set up a ratio; if 180 g of glucose contains 72 g of carbon, then 270 g glucose contains 108 g of carbon.

17. Calcium carbonate dissolves in acidic solutions as shown in the equation below.

$$CaCO_3(s) + 2\,H^+(aq) \rightarrow Ca^{2+}(aq) + H_2O(l) + CO_2(g)$$

If excess $CaCO_3$ is added to 0.250 liters of a 2.00-molar HNO_3 solution, what is the maximum volume of CO_2 gas that could be produced at standard temperature and pressure?

(A) 5.60 liters
(B) 11.2 liters
(C) 16.8 liters
(D) 33.6 liters

17. **A** H^+ is the limiting reagent, so we need to find out how many moles of H^+ are consumed.

$\text{Moles} = (\text{molarity})(\text{volume})$

$\text{Moles of } H^+ = (2.00\text{ M})(0.250\text{ L}) = 0.500\text{ moles}$

We can see from the balanced equation that for every 2 moles of H^+ consumed, 1 mole of CO_2 is produced, so if 0.500 moles of H^+ is consumed, 0.250 moles of CO_2 is produced.

$\text{Volume of gas at STP} =$
$(\text{Moles of gas})(22.4\text{ liters/mol})$

$\text{Volume of } CO_2 = (0.250\text{ moles})(22.4\text{ L/mol})$
$\qquad\qquad\qquad = 5.60\text{ liters}$

18. A monoprotic acid was titrated with a solution of NaOH. For 55.0 milliliters of the acid, 37.0 milliliters of a 0.450-molar solution of NaOH was required to reach the equivalence point. Which of the following expressions is equal to the initial concentration of the monoprotic acid?

(A) $\dfrac{(0.450)(0.037)}{(0.055)}\,M$

(B) $\dfrac{(0.450)(0.055)}{(0.037)}\,M$

(C) $\dfrac{(0.055)}{(0.450)(0.037)}\,M$

(D) $\dfrac{(0.037)}{(0.450)(0.055)}\,M$

18. **A** First let's find the number of moles of NaOH added.

$\text{Moles} = (\text{molarity})(\text{volume})$

$\text{Moles of NaOH} = (0.450)(0.037)\text{ moles}$

At the equivalence point, the moles of NaOH added are equal to the moles of acid originally present. Now we can find the original molar concentration of the acid.

$\text{Molarity} = \dfrac{\text{moles}}{\text{liters}}$

$\text{Molarity of the acid} = \dfrac{(0.450)(0.037)}{(0.055)}\,M$

QUESTIONS	EXPLANATIONS

19. $Zn(s) + NO_3^-(aq) + 10\ H^+(aq) \rightarrow$
$$Zn^{2+}(aq) + NH_4^+(aq) + 3\ H_2O(l)$$

Which of the following statements regarding the reaction shown above is correct?

(A) The oxidation number of hydrogen changes from +1 to 0.

(B) The oxidation number of hydrogen changes from +1 to –1.

(C) The oxidation number of nitrogen changes from +5 to –3.

(D) The oxidation number of nitrogen changes from +5 to +3.

19. C Remember that the oxidation state of oxygen is almost always –2. This means that the oxidation number of nitrogen in NO_3^- is +5. [Nitrogen + (–2)(3) = –1]

The oxidation state for hydrogen is almost always +1, so the oxidation state of nitrogen in NH_4^+ is –3. [Nitrogen + (1)(4) = 1] So the oxidation state of nitrogen changes from +5 to –3.

20. $HC_2H_3O_2(aq) + ClO^-(aq) \leftrightarrow HClO(aq) + C_2H_3O_2^-(aq)$

The standard free energy change for this reaction has a negative value. Based on this information, which of the following statements is true?

(A) K_a for $HC_2H_3O_2(aq)$ is less than K_a for $HClO(aq)$

(B) K_b for $C_2H_3O_2^-(aq)$ is less than K_b for $ClO^-(aq)$

(C) K_{eq} for the reaction is less than 1.

(D) The reaction occurs in the presence of a catalyst.

20. B If the standard free energy change is negative, then the forward reaction is favored under standard conditions, which means that K_{eq} will be greater than 1. For the forward reaction to be favored, $HC_2H_3O_2(aq)$ must be a stronger acid than $HClO(aq)$. If $HC_2H_3O_2(aq)$ is the stronger acid, then its conjugate, $C_2H_3O_2^-(aq)$, must be the weaker base. Thus, the base dissociation constant, K_b, is smaller for $C_2H_3O_2^-(aq)$ than for $ClO^-(aq)$.

21. A student added 0.20 mol of NaI and 0.40 mol of KI to 3 liters of water to create an aqueous solution. What is the minimum number of moles of $Pb(C_2H_3O_2)_2$ that the student must add to the solution to precipitate out all of the I^- ions as PbI_2?

(A) 2.40

(B) 1.20

(C) 0.60

(D) 0.30

21. D The solution contains 0.60 moles of I^- ions, 0.20 from NaI and 0.40 from KI. Each Pb^{2+} ion will remove 2 I^- ions, so the student needs to add only half as much $Pb(C_2H_3O_2)_2$ or 0.30 moles.

22. Which of the following statements regarding fluorine and nitrogen is NOT true?

(A) Fluorine has greater electronegativity.

(B) Fluorine has a greater first ionization energy.

(C) Fluorine has a greater atomic weight.

(D) Fluorine has a greater atomic radius.

22. D Fluorine has a smaller atomic radius than nitrogen. That's because fluorine has the same number of electron shells as nitrogen, but it has more protons in its nucleus. So each of fluorine's negatively charged electrons feels a stronger pull toward its more positively charged nucleus. This makes its radius smaller. All of the other answer choices are true.

QUESTIONS	EXPLANATIONS

23. Which of the statements below regarding elemental nitrogen is NOT true?

(A) It contains one sigma bond.
(B) It contains two pi bonds.
(C) It has a bond order of 3.
(D) It has a large dipole moment.

23. **D** Elemental nitrogen exists as a diatomic gas, N_2. N_2 is nonpolar, so it has a dipole moment of zero. N_2 has a triple bond, so it has a bond order of 3, with one sigma bond and two pi bonds.

24. The reaction of elemental chlorine with ozone in the atmosphere occurs by the two-step process shown below.

   I. $Cl + O_3 \rightarrow ClO + O_2$
   II. $ClO + O \rightarrow Cl + O_2$

Which of the statements below is true regarding this process?

(A) Cl is a catalyst.
(B) O_3 is a catalyst.
(C) ClO is a catalyst.
(D) O_2 is an intermediate.

24. **A** The overall reaction is as follows:

$$O_3 + O \rightarrow 2\,O_2$$

Cl acts as a catalyst in the reaction because it must be there for the reaction to occur, but it does not appear in the overall reaction. By the way, ClO is an intermediate in the process because it is the only component that's produced during the process and then completely consumed.

25. A student examined the line spectrum of a hydrogen atom and was able to conclude that when the electron in the hydrogen atom made the transition from the $n = 3$ to the $n = 1$ state, the frequency of the ultraviolet radiation emitted was 2.92×10^{15} Hz. When the same transition occurred in a He$^+$ ion, the frequency emitted was 1.17×10^{16} Hz. Which of the following best accounts for the difference?

(A) More energy is released by the He$^+$ transition because of the greater molar mass of the He$^+$ ion.
(B) More energy is released by the He$^+$ transition because of the greater nuclear charge of the He$^+$ ion.
(C) More energy is released by the He$^+$ transition because the ionic radius of He$^+$ is greater than the atomic radius of H.
(D) Less energy is released by the He$^+$ transition because the ionic radius of He$^+$ is smaller than the atomic radius of H.

25. **B** The frequency of radiation emitted by the He$^+$ is greater than the frequency emitted by the H atom for the same transition, so the energy emitted by the He$^+$ ion is greater (Remember: $E = hf$). That rules out (D).

The He$^+$ ion has 2 protons and the H atom has only one, so the electron is more strongly attracted to the He$^+$ ion and when it moves closer to the nucleus more energy will be released, so (B) is correct.

Choice (A) is wrong because the effect of mass in this situation is insignificant compared to charge.
Choice (C) is wrong because He$^+$ has a smaller radius than H.

QUESTIONS	EXPLANATIONS

26. $CH_3CH_2OH(g) + ...O_2(g) \rightarrow ...CO_2(g) + ...H_2O(g)$

The reaction above represents the oxidation of ethanol. How many moles of O_2 are required to oxidize 1 mole of CH_3CH_2OH?

(A) $\dfrac{3}{2}$ moles

(B) $\dfrac{5}{2}$ moles

(C) 3 moles

(D) $\dfrac{7}{2}$ moles

26. **C** Since we're given a coefficient here ($1CH_3CH_2OH$) and the answer choices look kind of ugly for backsolving, let's just put in the coefficient we have and work from there.

If we have 1 CH_3CH_2OH, then we have 2 Cs and thus, 2 CO_2s.

Also, if we have 1 CH_3CH_2OH, then we have 6 Hs, and thus, 3 H_2Os.

That means we have 7 Os on the right, so we must have 7 on the left. There is 1 O in CH_3CH_2OH, so we must get 6 from O_2, which means the coefficient of O_2 must be 3.

27. How many milliliters of water must be added to 10 milliliters of an HCl solution with a pH of 1 to produce a solution with a pH of 2?

(A) 10 ml
(B) 90 ml
(C) 100 ml
(D) 990 ml

27. **B** HCl is a strong acid, so it will completely dissociate into H^+ and Cl^- ions. A pH of 1 means $[H^+] = 0.1\ M$. A pH of 2 means $[H^+] = 0.01\ M$. So we need to go from a concentration of 0.1 M to 0.01 M. As the solution is diluted, the number of moles of H^+ will remain constant.

Moles = (molarity)(volume)

Moles of H^+ = $(0.1\ M)(10\ ml) = (0.01\ M)(x)$

$x = 100$ ml, but that's not the answer.

To get a 0.01-molar solution, we need 100 ml, but we already have 10 ml, so we need to *add* 90 ml of water. That's the answer.

28. Elemental iodine (I_2) is more soluble in carbon tetrachloride (CCl_4) than it is in water (H_2O). Which of the following statements is the best explanation for this?

(A) I_2 is closer in molecular weight to CCl_4 than it is to H_2O.
(B) The freezing point of I_2 is closer to that of CCl_4 than it is to that of H_2O.
(C) I_2 and CCl_4 are nonpolar molecules, while H_2O is a polar molecule.
(D) The heat of formation of I_2 is closer to that of CCl_4 than it is to that of H_2O.

28. **C** Like dissolves like. Polar molecules are more soluble in polar solvents, and nonpolar molecules are more soluble in nonpolar solvents.

QUESTIONS	EXPLANATIONS

29. A sample of water was electrolyzed to produce hydrogen and oxygen gas, as shown in the reaction below.

$$2\ H_2O(l) \rightarrow 2\ H_2(g) + O_2(g)$$

If 33.6 liters of gas were produced at STP, how many grams of water were consumed in the reaction?

(A) 11.1 grams
(B) 18.0 grams
(C) 24.0 grams
(D) 36.0 grams

29. B Moles of gas at STP = $\dfrac{\text{liters}}{22.4\ \text{moles/liter}}$

Moles of gas produced = $\dfrac{33.6\ \text{L}}{22.4\ \text{moles/L}}$ = 1.5 moles

3 moles of gas (1 oxygen and 2 hydrogen) are produced for every 2 moles of water consumed, so if 1.5 moles of gas were produced, then 1 mole of water was consumed.

Grams = (moles)(MW)

Grams of water consumed = (1 mole)(18.0 grams/mole) = 18.0 grams

30. The following data were gathered in an experiment to determine the density of a sample of an unknown substance.

Mass of the sample = 7.50 grams

Volume of the sample = 2.5 milliliters

The density of the sample should be reported as

(A) 3.00 grams per ml.
(B) 3.0 grams per ml.
(C) 3 grams per ml.
(D) 0.3 grams per ml.

30. B Density = $\dfrac{\text{mass}}{\text{volume}}$

Density of the sample = $\dfrac{7.50\ \text{grams}}{2.5\ \text{ml}}$ = 3.0 grams/ml

The volume (2.5 ml) has only two significant digits, so the final result (3.0 grams/ml) is limited to two significant digits.

31. $$2\ H_2O_2(g) + S(s) \rightarrow SO_2(g) + 2\ H_2O(g)$$

Based on the information given in the table below, what is the enthalpy change in the reaction represented above?

Substance	ΔH°_f (kJ/mol)
$H_2O_2(g)$	−150
$SO_2(g)$	−300
$H_2O(g)$	−250
$S(s)$	0

(A) −500 kJ
(B) −200 kJ
(C) 200 kJ
(D) 400 kJ

31. A We are given the heats of formation of all the reactants and products:

ΔH° for a reaction = $\Sigma(\Delta H^\circ_f$ for the products) − $\Sigma(\Delta H^\circ_f$ for the reactants)

First the products.

From SO_2, we get (−300 kJ) = −300 kJ

From H_2O, we get (2)(−250 kJ) = −500 kJ

So ΔH° for the products = (−300 kJ) + (−500 kJ) = −800 kJ

Now the reactants.

From H_2O_2, we get (2)(−150 kJ) = −300 kJ, and from S we get 0.

ΔH° for the reaction = (−800 kJ) − (−300 kJ) = (−800 kJ) + (300 kJ) = −500 kJ

QUESTIONS	EXPLANATIONS

32. A chemist creates a buffer solution by mixing equal volumes of a 0.2-molar HOCl solution and a 0.2-molar KOCl solution. Which of the following will occur when a small amount of KOH is added to the solution?

   I. The concentration of undissociated HOCl will increase.
  II. The concentration of OCl⁻ ions will increase.
  III. The concentration of H⁺ ions will increase.

  (A) I only
  (B) II only
  (C) III only
  (D) I and III only

32. B Look at the equilibrium reaction for this buffer solution.

$$HOCl(aq) \longleftrightarrow H^+(aq) + OCl^-(aq)$$

When OH^- is added, it will react to neutralize H^+. So adding OH^- is the same as removing H^+, and when H^+ is removed, the equilibrium shifts to the right to replace the lost H^+ ions. This causes the concentration of HOCl to decrease and the concentration of OCl^- to increase (that's choice II). The concentration of H^+ decreases a little bit but stays about the same, so really only choice II is correct.

33. A 15-gram sample of neon is placed in a sealed container with a constant volume of 8.5 liters. If the temperature of the container is 27°C, which of the following expressions gives the correct pressure of the gas? The ideal gas constant, R, is 0.08 (L-atm)/(mole-K).

A) $\dfrac{(15)(0.08)(300)}{(20)(8.5)}$ atm

(B) $\dfrac{(20)(0.08)(300)}{(15)(8.5)}$ atm

(C) $\dfrac{(20)(300)}{(15)(8.5)(0.08)}$ atm

(D) $\dfrac{(15)(300)}{(20)(8.5)(0.08)}$ atm

33. A First we need to find the number of moles of neon.

$$\text{Moles} = \frac{\text{grams}}{\text{MW}} = \frac{15 \text{ g}}{20 \text{ g/mol}} = n$$

Now we can use the ideal gas equation to find the pressure. Don't forget to convert 27°C to 300 K.

$$P = \frac{nRT}{V} =$$

$$\frac{\left(\frac{15}{20} \text{ mol}\right)\left(0.08 \frac{\text{L-atm}}{\text{mol-K}}\right)(300 \text{ K})}{(8.5 \text{ L})} =$$

$$\frac{(15)(0.08)(300)}{(20)(8.5)} \text{ atm}$$

QUESTIONS	EXPLANATIONS

34.

$$C(s) + CO_2(g) + energy \longleftrightarrow 2\ CO(g)$$

The system above is currently at equilibrium in a closed container. Which of the following changes to the system would serve to increase the number of moles of CO present at equilibrium?

   I. Raising the temperature
  II. Increasing the volume of the container
 III. Adding more C(s) to the container

(A) I only
(B) II only
(C) I and II only
(D) I and III only

34. C Use Le Châtelier's law.

I. Raising the temperature favors the endothermic, forward reaction, increasing the CO produced at equilibrium, so (I) is correct.

II. Increasing the volume of the container favors the side with more moles of gas. That's the CO side, so (II) is correct.

III. Adding more of a solid will have no effect on an equilibrium situation, so choice (III) is not correct.

35. In a titration experiment, a 0.10-molar $H_2C_2O_4$ solution was completely neutralized by the addition of a 0.10-molar NaOH solution. Which of the diagrams below illustrates the change in pH that accompanied this process?

35. D The diagram for the titration of a diprotic acid (such as $H_2C_2O_4$) by a base shows increasing pH and two endpoints.

(A)

(C)

(B)

(D)

QUESTIONS	EXPLANATIONS

36. If 52 grams of $Ba(NO_3)_2$ (molar mass 260 grams) are completely dissolved in 500 milliliters of distilled water, what are the concentrations of the barium and nitrate ions?

    (A) $[Ba^{2+}] = 0.10\ M$ and $[NO_3^-] = 0.10\ M$
    (B) $[Ba^{2+}] = 0.20\ M$ and $[NO_3^-] = 0.40\ M$
    (C) $[Ba^{2+}] = 0.40\ M$ and $[NO_3^-] = 0.40\ M$
    (D) $[Ba^{2+}] = 0.40\ M$ and $[NO_3^-] = 0.80\ M$

36. D First let's find the number of moles of $Ba(NO_3)_2$, Ba^{2+}, and NO_3^- in the solution.

$$\text{Moles} = \frac{\text{grams}}{\text{MW}}$$

$$\text{Moles of } Ba(NO_3)_2 = \frac{52\ g}{260\ g/mol} = 0.20\ \text{moles}$$

For every mole of $Ba(NO_3)_2$ in solution, we get 1 mole of Ba^{2+} and 2 moles of NO_3^-, so we have 0.20 moles of Ba^{2+} and 0.04 moles of NO_3^-.

Now we can find the concentrations of Ba^{2+} and NO_3^-.

$$\text{Molarity} = \frac{\text{moles}}{\text{liters}}$$

$$[Ba^{2+}] = \frac{0.20\ \text{moles}}{0.500\ L} = 0.40\ M$$

$$[NO_3^-] = \frac{0.40\ \text{moles}}{0.500\ L} = 0.80\ M$$

37. All of the following statements concerning the alkali metals are true EXCEPT

    (A) They are strong oxidizing agents.
    (B) They form ions with a +1 oxidation state.
    (C) As the atomic numbers of the alkali metals increase, their electronegativity decreases.
    (D) As the atomic numbers of the alkali metals increase, the first ionization energy decreases.

37. A The alkali metals (Li, Na, K, and others) are not strong oxidizing agents, they are strong *reducing* agents. They give up their single valence electrons easily, so they are easily oxidized (LEO). If they are easily oxidized, they are strong reducing agents.

38. As a beaker of water is heated over a flame, the temperature increases steadily until it reaches 373 K. At that point, the beaker is left on the open flame, but the temperature remains at 373 K as long as water remains in the beaker. This is because at 373 K, the energy supplied by the flame

    (A) is completely absorbed by the glass beaker.
    (B) is less than the energy lost by the water through electromagnetic radiation.
    (C) is dissipated by the water as visible light.
    (D) is used to overcome the heat of vaporization of the water.

38. D The boiling point of water is 373 K (100°C). At this point, the energy put into the water is used to overcome the intermolecular forces of attraction of the water molecules as it vaporizes. (That's the heat of vaporization.)

39. For which of the following reactions will the equilibrium constants K_c and K_p have the same value?

    (A) $2\,N_2O_5(g) \leftrightarrow 2\,NO_2(g) + O_2(g)$
    (B) $2\,CO_2(g) \leftrightarrow 2\,CO(g) + O_2(g)$
    (C) $H_2O\,(g) + CO(g) \leftrightarrow H_2(g) + CO_2(g)$
    (D) $3\,O_2(g) \leftrightarrow 2\,O_3(g)$

39. **C** The expression that connects K_p and K_c is $K_p = K_c\,(RT)^n$, where n is the change in moles of gas over the course of the reaction. For choice (C), there are 2 moles of reactant gases and 2 moles of product gases, so $n = 0$ and $K_p = K_c$. None of the other choices has $n = 0$.

40. Each of the following compounds was added to distilled water at 25°C. Which one produced a solution with a pH that was less than 7?

    (A) O_2
    (B) NaI
    (C) MgO
    (D) SO_2

40. **D** SO_2 is an acid anhydride; it reacts with water to produce H_2SO_3. So when SO_2 is added to water, the solution will be acidic and its pH will be less than 7. N_2 and O_2 will not react with water, NaI is a neutral salt, and MgO is a basic anhydride that will produce a solution with a pH greater than 7.

41. ...Be_2C + ...H_2O → ...$Be(OH)_2$ + ...CH_4

    When the equation for the reaction above is balanced with the lowest whole-number coefficients, the coefficient for H_2O will be

    (A) 1
    (B) 2
    (C) 3
    (D) 4

41. **D** Backsolve, starting with (C).

    If there are 3 H_2Os, there is no way to balance O on the other side of the equation because, with a whole number coefficient for $Be(OH)_2$, there can be only an even number of Os. That means that (A) is also wrong.

    Try (D).

    If there are 4 H_2Os, then there are 2 $Be(OH)_2$s.

    If there are 2 $Be(OH)_2$s, then there is 1 Be_2C.

    If there is 1 Be_2C, then there is 1 CH_4.

    Now there are 8 Hs on the left and 8 Hs on the right, so the equation is balanced with the lowest whole number coefficients and (D) is correct.

42. A sample of an ideal gas is placed in a sealed container of constant volume. If the temperature of the gas is increased from 40°C to 70°C, which of the following values for the gas will NOT increase?

    (A) The average speed of the molecules of the gas
    (B) The average kinetic energy of the molecules of the gas
    (C) The pressure exerted by the gas
    (D) The density of the gas

42. **D** According to kinetic-molecular theory, as the temperature of a gas is increased, the speed, kinetic energy, and frequency of collision of its molecules all increase. From the ideal gas law, we know that as temperature increases at constant volume, pressure increases. The temperature increase affects neither the volume of the container nor the mass of the gas contained, so the density of the gas is not changed.

43. An $^{222}_{86}Rn$ nuclide decays through the emission of two beta particles and two alpha particles. The resulting nuclide is

(A) $^{214}_{84}Po$

(B) $^{210}_{84}Po$

(C) $^{214}_{83}Bi$

(D) $^{210}_{83}Bi$

43. **A** Let's do the math.

$$^{222}_{86}Rn - {}_{-1}\beta - {}_{-1}\beta - {}^{4}_{2}\alpha - {}^{4}_{2}\alpha =$$

For the mass number we have

$$222 - 4 - 4 = 214$$

For the proton number we have

$$86 - (-1) - (-1) - 2 - 2 = 86 + 2 - 4 = 84$$

So the answer is $^{214}_{84}Po$.

44. When 80.0 ml of a 0.40 M NaI solution is combined with 20.0 ml of a 0.30 M CaI$_2$ solution, what will be the molar concentration of I$^-$ ions in the solution?

(A) 0.70 M

(B) 0.44 M

(C) 0.38 M

(D) 0.35 M

44. **B** The I$^-$ ions from the two salts will both be present in the solution, so we need to find the number of moles of I$^-$ contributed by each salt.

Moles = (molarity)(volume)

Each NaI produces 1 I$^-$

Moles of I$^-$ from NaI = (0.40 M)(0.080 L)
= 0.032 mole

Each CaI$_2$ produces 2 I$^-$

Moles of I$^-$ from CaI$_2$ = (2)(0.30 M)(0.020 L)
= 0.012 mole

To find the number of moles of Cl$^-$ in the solution, add the two together.

0.032 mole + 0.012 mole = 0.044 mole

Now use the formula for molarity to find the concentration of I$^-$ ions. Don't forget to add the volumes of the two solutions.

Molarity = $\dfrac{moles}{liters} = \dfrac{(0.044\,mol)}{(0.100\,L)}$ = 0.44 M

45. Hydrogen sulfide is a toxic waste product of some industrial processes. It can be recognized by its distinctive rotten egg smell. One method for elimininating hydrogen sulfide is by reaction with dissolved oxygen, as shown below.

$$2 H_2S + O_2 \rightarrow 2 S + 2 H_2O$$

If 102 grams of H_2S are combined with 64 grams of O_2, what is the maximum mass of elemental sulfur that could be produced by the reaction?

(A) 32 grams
(B) 48 grams
(C) 64 grams
(D) 96 grams

45. **D** We need to find the limiting reagent.

$$Moles = \frac{grams}{MW}$$

$$Moles \text{ of } H_2S = \frac{102 \text{ grams}}{34 \text{ g/mol}} = 3 \text{ moles}$$

$$Moles \text{ of } O_2 = \frac{64 \text{ grams}}{32 \text{ g/mol}} = 2 \text{ moles}$$

We can see from the balanced equation that 2 moles of H_2S are consumed for every 1 mole of O_2. Since the number of moles of H_2S is less than twice the number of moles of O_2, H_2S will run out first and is the limiting reagent.

Also, from the balanced equation, the number of moles of S produced is equal to the number of moles of H_2S consumed, so 3 moles of S are produced.

Grams = (moles)(MW)

Grams of S = (3 moles)(32 g/mol) = 96 grams

46. $$H_2O(g) \rightarrow H_2O(l)$$

Which of the following is true of the values of ΔH, ΔS, and ΔG for the reaction shown above at 25°C?

	ΔH	ΔS	ΔG
(A)	Positive	Positive	Positive
(B)	Positive	Negative	Negative
(C)	Negative	Negative	Positive
(D)	Negative	Negative	Negative

46. **D** Heat is given off when water condenses, so ΔH is negative. The entropy decreases in a phase change from gas to liquid, so ΔS is negative. The water condenses spontaneously at room temperature, so ΔG is negative.

47. A student added 1 liter of a 1.0 M Na_2SO_4 solution to 1 liter of a 1.0 M $Ag(C_2H_3O_2)$ solution. A silver sulfate precipitate formed, and nearly all of the silver ions disappeared from the solution. Which of the following lists the ions remaining in the solution in order of decreasing concentration?

(A) $[SO_4^{2-}] > [C_2H_3O_2^-] > [Na^+]$
(B) $[C_2H_3O_2^-] > [Na^+] > [SO_4^{2-}]$
(C) $[Na^+] > [SO_4^{2-}] > [C_2H_3O_2^-]$
(D) $[Na^+] > [C_2H_3O_2^-] > [SO_4^{2-}]$

47. **D** At the start, the concentrations of the ions are as follows:

$[Na^+] = 2 M$

$[SO_4^{2-}] = 1 M$

$[Ag^+] = 1 M$

$[C_2H_3O_2^-] = 1 M$

After Ag_2SO_4 forms, the concentrations are

$[Na^+] = 2 M$

$[SO_4^{2-}] = 0.5 M$

$[Ag^+] = 0 M$

$[C_2H_3O_2^-] = 1 M$

So from greatest to least

$[Na^+] > [C_2H_3O_2^-] > [SO_4^{2-}]$

QUESTIONS	EXPLANATIONS

48. Iron ore is converted to pure iron by the following reaction, which occurs at extremely high temperatures:

$$Fe_2O_3 + 3\,CO \rightarrow 2\,Fe + 3\,CO_2$$

A 1,600-gram sample of iron ore reacted completely to form 558 grams of pure iron. What was the percent of Fe_2O_3 (molecular weight 160) by mass in the original sample?

(A) 25%
(B) 33%
(C) 50%
(D) 100%

48. C $Moles = \dfrac{grams}{MW}$

$Moles\ of\ Fe = \dfrac{558\ grams}{55.8\ g/mol} = 10\ moles$

2 moles of Fe are produced for every 1 mole of Fe_2O_3 consumed, so if 10 moles of Fe were produced, 5 moles of Fe_2O_3 were consumed.

$Grams = (moles)(MW)$

$Grams\ of\ Fe_2O_3 = (5\ moles)(160\ g/mol)$
$\qquad\qquad\qquad = 800\ grams$

Percent of Fe_2O_3 by

$mass = \dfrac{grams\ of\ Fe_2O_3}{grams\ of\ sample} = \dfrac{800\ g}{1600\ g} = 0.50 = 50\%$

49. Gold and silver have been used throughout history to make coins. Which of the following statements could account for the popularity of these metals for use in coins?

(A) Network bonds make these metals especially durable.
(B) Negative oxidation potentials for these metals make them especially unreactive with their surroundings.
(C) These metals are among the lightest of the elements.
(D) These metals are commonly found in nature.

49. B A coin must be stable; it must not react with its surroundings. Elemental silver and gold are among the few metals that have negative oxidation potentials, which makes them unreactive. All of the statements in the other answer choices are false.

50.
$$2\,HI(g) \rightarrow H_2(g) + I_2(g)$$

For the reaction given above, $H°$ is –50 kJ. Based on the information given in the table below, what is the average bond energy of the H–I bond?

Bond	Average Bond Energy (kJ/mol)
H–H	440
I–I	150

(A) 270 kJ/mol
(B) 540 kJ/mol
(C) 590 kJ/mol
(D) 640 kJ/mol

50. A Use the following equation to solve the problem.

$H° =$ [Bond energies of bonds broken] – [Bond energies of the bonds formed]

$H° = [(2)(H–I)] – [(H–H) + (I–I)]$

$–50kJ = (2)(H–I) – [440\ kJ + 150\ kJ]$

$(2)(H–I) = –50\ kJ + 590\ kJ$

$(2)(H–I) = 540\ kJ$

$H–I = 270\ kJ$

QUESTIONS	EXPLANATIONS

QUESTIONS

51. All of the elements listed below are gases at room temperature. Which gas would be expected to show the greatest deviation from ideal behavior?

  (A) He
  (B) Ar
  (C) Kr
  (D) Xe

Questions 52 and 53 refer to the information given below.

$$F_2(g) + 2\,ClO_2(g) \rightarrow 2\,FClO_2(g)$$

Experiment	$[F_2]$ (M)	$[ClO_2]$ (M)	Initial rate of disappearance of F_2 (M/sec)
1	0.10	0.010	1.2×10^{-3}
2	0.20	0.010	2.4×10^{-3}
3	0.40	0.020	9.6×10^{-3}

52. Based on the data given in the table, which of the following expressions is equal to the rate law for the reaction given above?
  (A) Rate = $k[F_2]$
  (B) Rate = $k[ClO_2]$
  (C) Rate = $k[F_2][ClO_2]$
  (D) Rate = $k[F_2]^2[ClO_2]$

53. What is the initial rate of disappearance of ClO_2 in experiment 2?

  (A) 1.2×10^{-3} M/sec
  (B) 2.4×10^{-3} M/sec
  (C) 4.8×10^{-3} M/sec
  (D) 7.0×10^{-3} M/sec

EXPLANATIONS

51. **D** Deviation from ideal behavior comes about because of attractive forces between gas molecules and because of molecular size. Of the noble gases listed, Xe has the most electrons, so it will have the strongest London dispersion forces, which are the only attractions experienced by the noble gases. Xe is also the largest atom listed.

52. **C** Compare experiments 1 and 2. When $[F_2]$ is doubled while $[ClO_2]$ is held constant, the rate doubles, so the reaction is first order with respect to $[F_2]$.

The next part is a little tricky. Compare experiments 1 and 3. When both $[F_2]$ and $[ClO_2]$ are doubled, the rate quadruples. (Quadrupling is like doubling twice.) One of the rate doublings came from $[F_2]$, so the other one must have come from $[ClO_2]$. So doubling $[ClO_2]$ also causes the rate to double, making the reaction first order with respect to $[ClO_2]$.

The reaction is first order with respect to $[F_2]$ and first order with respect to $[ClO_2]$, so the rate law says that Rate = $k[F_2][ClO_2]$.

53. **C** We can see from the balanced equation that for every mole of F_2 consumed, 2 moles of ClO_2 are consumed. So ClO_2 will disappear at twice the rate that F_2 disappears. In experiment 2, F_2 disappears at the rate of 2.4×10^{-3} M/sec, so ClO_2 will disappear at twice that rate, or 4.8×10^{-3} M/sec.

QUESTIONS	EXPLANATIONS

54. Hypobromous acid, HBrO, is added to distilled water. If the acid dissociation constant for HBrO is equal to 2×10^{-9}, what is the concentration of HBrO when the pH of the solution is equal to 5?

(A) 5-molar
(B) 1-molar
(C) 0.1-molar
(D) 0.05-molar

54. D Use the K_a expression for HBrO. In the denominator, we can ignore the amount of HBrO that dissociates because it is such a small number.

$$K_a = \frac{\left[H^+\right]\left[BrO^-\right]}{\left[HBrO\right]}$$

If the pH is 5, then $[H^+] = [BrO^-] = 1^{-5} M$

$$2 \times 10^{-9} = \frac{\left(1 \times 10^{-5}\right)\left(1 \times 10^{-5}\right)}{\left[HBrO\right]}$$

$$[HBrO] = \frac{\left(1 \times 10^{-10}\right)}{\left(2 \times 10^{-9}\right)} M = 0.5^{-1} M = 0.05\ M$$

55. At 25°C, the vapor pressure of water is 24 mmHg. Which of the following expressions gives the vapor pressure of a solution created by adding 2.0 moles of sucrose to 55 moles of water?

(A) $\dfrac{(24)(2.0)}{(55)}$ mmHg

(B) $\dfrac{(2.0)}{(24)(55)}$ mmHg

(C) $\dfrac{(24)(55)}{(57)}$ mmHg

(D) $\dfrac{(55)}{(24)(57)}$ mmHg

55. C Use Raoult's law.

$$VP_{SOLUTION} = (VP_{SOLVENT})(\text{mole fraction of solvent})$$

$$VP_{SOLUTION} =$$

$$\frac{55 \text{ moles of water}}{(55 + 2.0) \text{ moles of solution}} = (24)\frac{55}{57} = 24 \text{ mmHg}$$

$$VP_{SOLUTION} = \frac{(24)(55)}{(57)} \text{ mmHg}$$

56.
$$PCl_5(g) \leftrightarrow PCl_3(g) + Cl_2(g)$$

When PCl_5 is placed in an evacuated container at 250°C, the above reaction takes place. The pressure in the container before the reaction takes place is 6 atm and is entirely due to the PCl_5 gas. After the reaction comes to equilibrium, the partial pressure due to Cl_2 is found to be 2 atm. What is the value of the equilibrium constant, K_p, for this reaction?

(A) 0.5
(B) 1
(C) 2
(D) 5

56. B First we need to find out the equilibrium partial pressures of each of the gases. The question says that the partial pressure of Cl_2 at equilibrium is 2 atm.

From the balanced equation, for every mole of Cl_2 generated by the reaction, 1 mole of PCl_3 is also generated. There was no Cl_2 or PCl_3 to start with, so if there are 2 atm of Cl_2 at equilibrium, there must also be 2 atm of PCl_3.

Also from the balanced equation, for every mole of Cl_2 generated, 1 mole of PCl_5 must be consumed. So if there are 2 moles of Cl_2 at equilibrium, there must be $6 - 2 = 4$ atm of PCl_5.

Now we can use the K_p expression.

$$K_p = \frac{\left(P_{PCl_3}\right)\left(P_{Cl_2}\right)}{\left(P_{PCl_5}\right)} = \frac{(2)(2)}{(4)} = 1$$

QUESTIONS	EXPLANATIONS

57. $Co^{2+} + 2\,e^- \rightarrow Co$ $\quad E^\circ = -0.28$ V

$Cr^{3+} + e^- \rightarrow Cr^{2+}$ $\quad E^\circ = -0.41$ V

According to the information shown above, which of the following statements is true regarding the reaction below?

$$2\,Co + Cr^{3+} \rightarrow 2\,Co^{2+} + Cr^{2+}$$

(A) Cr^{3+} acts as the reducing agent.
(B) The reaction would take place in a galvanic cell.
(C) K_{eq} for the reaction is greater than 1.
(D) Cobalt metal is reduced in the reaction.

57. C For a spontaneous reaction, G° is negative.

From the equation $G^\circ = -nFE^\circ$, if G° is negative, E° must be positive.

From the equation $G^\circ = -RT\ln K$, if G° is negative, $\ln K$ must be positive, so K must be greater than 1.

58. What is the pH of a solution made by mixing 200 milliliters of a 0.20-molar solution of NH_3 with 200 milliliters of a 0.20 milliliters of an NH_4Cl solution? (The base dissociation constant, K_b, for NH_3 is 1.8×10^{-5}.)

(A) Between 4 and 5
(B) Between 5 and 6
(C) Between 8 and 9
(D) Between 9 and 10

58. D This solution is a buffer with equal concentrations of acid (NH_4^+) and conjugate base (NH_3). For a buffer with equal concentrations of acid and conjugate base, $pH = pK_a$ and $pOH = pK_b$.

We are given K_b, so we can estimate pOH. $K_b = 1.8 \times 10^{-5}$, so pK_b and pOH are between 4 and 5 (actually it's 4.7). But we need the pH.

pH + pOH = 14

pH + (between 4 and 5) = 14

pH is between 9 and 10

59. $$POCl_3(l) \rightarrow POCl_3(g)$$

For the reaction above, H is 50 kilojoules per mole and S is 100 joules per mole. What is the boiling point of $POCl_3$? (Assume that H and S remain constant with changing temperature.)

(A) 2 K
(B) 50 K
(C) 200 K
(D) 500 K

59. D At the boiling point, the reaction is at equilibrium. Remember that, at equilibrium, $G = 0$. Use the relationship $G = H - TS$ and don't forget to convert kilojoules into joules.

$G = H - TS$

$0 = 50{,}000 - (T)(100)$

$T = \dfrac{50{,}000}{100}$ K $= 500$ K

60. When fluorine gas is bubbled through a concentrated aqueous solution of potassium chloride at room temperature, which of the following would be expected to occur?

   (A) A reaction will occur, and chlorine gas will be produced.
   (B) A reaction will occur, and potassium fluoride will precipitate.
   (C) A reaction will occur, and solid chlorine will precipitate.
   (D) A reaction will occur, and solid potassium will precipitate.

60. **A** Fluorine gas is an extremely strong oxidizing agent. When F_2 gas is bubbled through a solution containing any other halogen anions, F^- ions will replace the other halogen ions in the solution. Chlorine ions can be oxidized to form Cl_2 gas, which will bubble out of the solution.

QUESTIONS	EXPLANATIONS

Section II—Free Response

1. At 25°C the solubility product consant, K_{sp}, for nickel hydroxide, $Ni(OH)_2$, is 1.6×10^{-14}.

(a) Write a balanced equation for the solubility equilibrium for $Ni(OH)_2$ in water.

(a) $Ni(OH)_2 \leftrightarrow Ni^{2+} + 2\ OH^-$

(b) What is the molar solubility of $Ni(OH)_2$ in pure water at 25°C?

(b) Use the following K_{sp} expression: $K_{sp} = [Ni^{2+}][OH^-]^2$

One Ni^{2+} is produced for every $Ni(OH)_2$ that dissolves, so the molar solubility of $Ni(OH)_2$ will be the same as the concentration of Ni^{2+}. Two OH^-s will be produced for every Ni^{2+}, so $[OH^-]$ will be twice as big as $[Ni^{2+}]$.

Let $x = [Ni^{2+}]$

$K_{sp} = [Ni^{2+}][OH^-]^2$

$1.6 \times 10^{-14} = (x)(2x)^2 = 4x^3$

$x = 1.6\ 10^{-5}\ M$

So the solubility of $Ni(OH)_2$ at 25°C is 1.6×10^{-5} moles/liter.

(c) A 1.0-molar NaOH solution is slowly added to a saturated solution of $Ni(OH)_2$ at 25°C. If excess solid $Ni(OH)_2$ remains in the solution throughout the procedure, what is the concentration of Ni^{2+} ions in the solution at the moment when the pH is equal to 11?

(c) Use the following K_{sp} expression again: $K_{sp} = [Ni^{2+}][OH^-]^2$

First let's find $[OH^-]$.

If the pH is 11, then from pH + pOH = 14, we know that the pOH is 3.

If the pOH is 3, then from pOH = –log $[OH^-]$, we know that $[OH^-] = 1.0\ 10^{-3}$.

Now we can use the K_{sp} expression to find $[Ni^{2+}]$.

$K_{sp} = [Ni^{2+}][OH^-]^2$

$1.6 \times 10^{-14} = [Ni^{2+}](1.0 \times 10^{-3})^2$

$1.6 \times 10^{-14} = [Ni^{2+}](1.0 \times 10^{-6})$

$[Ni^{2+}] = 1.6 \times 10^{-8}\ M$

QUESTIONS	EXPLANATIONS
(d) Predict whether a precipitate will form when 200.0 milliliters of a 5.0×10^{-5}-molar KOH solution is mixed with 300.0 milliliters of a 2.0×10^{-4}-molar $Ni(NO_3)_2$ solution at 25°C. Show calculations to support your prediction.	(d) First we need to find the concentrations of the Ni^{2+} and OH^- ions.

Moles = (molarity)(volume)

Moles of $OH^- = (5.0 \times 10^{-5} M)(0.200 L)$
$= 1.0 \times 10^{-5}$ mol

Moles of $Ni^{2+} = (2.0 \times 10^{-4})(0.300 L)$
$= 6.0 \times 10^{-5}$ mol

Remember to add the two volumes:
$(0.300 L) + (0.200 L) = 0.500 L$

$$\text{Molarity} = \frac{\text{moles}}{\text{liters}}$$

$$[Ni^{2+}] = \frac{\left(6.0 \times 10^{-5} \text{ mol}\right)}{\left(0.500 \text{ L}\right)} = 1.2 \times 10^{-4} M$$

$$[OH^-] = \frac{\left(1.0 \times 10^{-5} \text{ mol}\right)}{\left(0.500 \text{ L}\right)} = 2.0 \times 10^{-5} M$$

Now test the solubility expression using the initial values, to find the reaction quotient.

$Q = [Ni^{2+}][OH^-]^2$

$Q = (1.2 \times 10^{-4})(2.0 \times 10^{-5})^2 = 4.8 \times 10^{-14}$

Q is greater than K_{sp}, so a precipitate will form.

(e) At 25°C, 100 ml of a saturated $Ni(OH)_2$ solution was prepared.

   (i) Calculate the mass of $Ni(OH)_2$ present in the solution.

   (ii) Calculate the pH of the solution.

   (iii) If the solution is allowed to evaporate to a final volume of 50 ml, what will be the pH? Justify your answer.

(e) (i) Use the result from (b). First convert moles/L to grams/L. The formula weight of $Ni(OH)_2$ is 92.69 g/mol.

Grams/L = (moles/L)(MW) = $(1.6 \times 10^{-5}$ mol/L) (92.69 g/mol) = 0.0015 g/L

Now you can find the grams in 0.100 L. (0.0015 g/L)(0.100 L) = 0.00015 grams.

(ii) From (b), $[OH^-]$ will be twice as large as $[Ni^{2+}]$. So $[OH^-] = 3.2 \times 10^{-5} M$

pOH = $-\log[OH^-] = -\log(3.2 \times 10^{-5}) = 4.49$

pH = 14 − pOH = 14 − 4.49 = 9.51

(iii) Since the solution is saturated, evaporation will not change the concentrations, so the pH will remain 9.51.

QUESTIONS	EXPLANATIONS

2. In two separate experiments, a sample of an unknown hydrocarbon was burned in air, and a sample of the same hydrocarbon was placed into an organic solvent.

(a) When the hydrocarbon sample was burned in a reaction that went to completion, 2.2 grams of water and 3.6 liters of carbon dioxide were produced under standard conditions. What is the empirical formula of the hydrocarbon?

(a) All of the hydrogen in the water and all of the carbon in the carbon dioxide must have come from the hydrocarbon.

First we'll find out how many moles of hydrogen there are.

$$\text{Moles} = \frac{\text{grams}}{\text{MW}}$$

$$\text{Moles of } H_2O = \frac{(2.2 \text{ g})}{(18.0 \text{ g/mol})} = 0.12 \text{ moles}$$

Every mole of water contains two moles of hydrogen, so there are 0.24 moles of hydrogen.

Now let's find out how many moles of carbon dioxide there are.

$$\text{Moles} = \frac{\text{liters}}{22.4 \text{ L/mol}}$$

$$\text{Moles of } CO_2 = \frac{(3.6 \text{ L})}{(22.4 \text{ L/mol})} = 0.16 \text{ moles}$$

Every mole of CO_2 contains one mole of carbon, so there are 0.16 moles of carbon.

The ratio of H to C is $\dfrac{0.24}{0.16} = \dfrac{1.5}{1} = \dfrac{3}{2}$

So the empirical formula of the hydrocarbon is C_2H_3.

(b) When 4.05 grams of the unknown hydrocarbon was placed in 100.0 grams of benzene, C_6H_6, the freezing point of the solution was measured to be 1.66°C. The normal freezing point of benzene is 5.50°C and the freezing point depression constant for benzene is 5.12°C/m. What is the molecular weight of the unknown hydrocarbon?

(b) First we'll find the molality of the solution. The freezing point depression, T, is

$$5.50°C - 1.66°C = 3.84°C.$$

$$T = kmx$$

$x = 1$ because hydrocarbons don't dissociate in solution. Now solve for m.

$$m = \frac{\Delta T}{k} = \frac{(3.84°C)}{(5.12°C/m)} = 0.750\ m$$

From the molality of the solution, we can find the number of moles of the unknown hydrocarbon.

$$\text{Molality} = \frac{\text{moles of solute}}{\text{kg of solvent}}$$

Solve for moles.

Moles = (molality)(kg of solvent)

Moles of hydrocarbon = (0.750 m)(0.100 kg)
$$= 0.0750\ \text{moles}$$

Now we can find the molecular weight of the hydrocarbon.

$$\text{MW} = \frac{\text{grams}}{\text{moles}} = \frac{(4.05\ \text{g})}{(0.0750\ \text{mol})} = 54.0\ \text{g/mol}$$

(c) What is the molecular formula and name of the hydrocarbon?

(c) From (a) we know that the empirical formula is C_2H_3. For C_2H_3, the molecular weight would be 27 g/mol.

From (b) we know that the molecular weight of the compound is 54 g/mol. That's twice as large as the weight for the empirical formula, so we can just double the numbers in the empirical formula to get the molecular formula, C_4H_6, butyne.

(d) Write the balanced equation for the combustion reaction that took place in (a).

(d) $2\ C_4H_6 + 11\ O_2 \rightarrow 6\ H_2O + 8\ CO_2$

(e) Draw two isomers for the hydrocarbon.

(e)

1-butyne   2-butyne

3. A strip of Ni metal is placed in a 1-molar solution of $Ni(NO_3)_2$ and a strip of Ag metal is placed in a 1-molar solution of $AgNO_3$. An electrochemical cell is created when the two solutions are connected by a salt bridge and the two metal strips are connected by wires to a voltmeter.

(a) Write the balanced chemical equation for the overall reaction that occurs in the cell, and calculate the cell potential, E^o.

(a) The half-reactions in the cell are as follows:

$$Ni \rightarrow Ni^{2+} + 2\,e^- \qquad E^o = +0.25\text{ V}$$

$$Ag^+ + e^- \rightarrow Ag \qquad E^o = +0.80\text{ V}$$

You can get the overall reaction by adding the half-reactions and balancing the electrons exchanged. You get the reaction potential by adding the potentials for the half-reactions and ignoring the coefficients in the balanced equation.

$$Ni + 2\,Ag^+ \rightarrow Ni^{2+} + 2\,Ag \qquad E^o = +1.05\text{ V}$$

(b) Calculate how many grams of metal will be deposited on the cathode if the cell is allowed to run at a constant current of 1.5 amperes for 8.00 minutes.

(b) At the cathode, Ag^+ ions are reduced to $Ag(s)$, so the metal deposited at the cathode will be solid silver.

First we need to find out how many electrons were supplied by the current.

Coulombs = (amperes)(seconds)

Coulombs = (1.50 A)(8.00 min)(60 sec/min)
$\qquad$ = 720 C

$$\text{Moles of electrons} = \frac{\text{coulombs}}{96{,}500} = \frac{720}{96{,}500}$$
$$= 0.00746 \text{ moles}$$

From the silver half-reaction, we know that for every mole of electrons consumed, one mole of $Ag(s)$ is produced. So we have 0.00746 moles of $Ag(s)$.

Grams = (moles)(MW)

Grams of $Ag(s)$ = (0.00746 mol)(107.9 g/mol)
$\qquad$ = 0.805 grams

QUESTIONS	EXPLANATIONS

(c) Calculate the value of the standard free energy change, $\Delta G°$, for the cell reaction.

(c) Use the following expression:

$$\Delta G° = -nFE°$$

$n = 2$, because two moles of electrons are exchanged in the redox reaction.

$E° = 1.05$ V, from part (a)

$F = 96,500$ C/mol

$$\Delta G° = -(2)(96,500 \text{ C/mol})(1.05 \text{ V})$$
$$= -202,650 \text{ C-V/mol} = -202,650 \text{ J/mol}$$
$$= -203 \text{ kJ/mol}$$

(d) Calculate the cell potential, E, at 25°C for the cell shown above if the initial concentration of $Ni(NO_3)_2$ is 0.100-molar and the initial concentration of $AgNO_3$ is 1.20-molar.

(d) Use the Nernst equation for 25°C:

$$E_{cell} = E° - \frac{0.0592}{n} \log Q$$

$$E_{cell} = E° - \frac{0.0592}{n} \log \frac{\left[Ni^{2+}\right]}{\left[Ag^+\right]^2}$$

$$E_{cell} = 1.05 \text{ V} - \frac{0.0592}{2} \log \frac{(0.100)}{(1.20)} \text{ V}$$
$$E_{cell} = 1.05 \text{ V} - (-0.034 \text{ V}) = 1.08 \text{ V}$$

(e) Is the reaction in the cell spontaneous under the conditions described in part (d)? Justify your answer.

(e) The cell potential is positive, so the reaction is spontaneous.

QUESTIONS	EXPLANATIONS

4. You will be given three chemical reactions below. In part (i), write the balanced equation for the reaction, leaving coefficients in terms of lowest whole numbers. Then answer the question pertaining to that reaction in part (ii). For each of the following three reactions, assume that solutions are aqueous unless it says otherwise. Substances in solutions should be represented as ions if these substances are extensively ionized. Omit formulas for ions or molecules that are not affected by the reaction. Only equations inside the answer boxes will be graded.

(a) Solutions of sodium acetate and nitric acid are mixed.

(i) Balanced equation:

(i) $H^+ + C_2H_3O_2^- \Rightarrow HC_2H_3O_2$

(ii) Identify the spectator ion in this reaction.

(ii) Nitrate (NO_3^-) is not included in the balanced equation, so it is the spectator ion.

(b) A piece of solid calcium is heated in oxygen gas.

(i) Balanced equation:

(i) $2\,Ca + O_2 \Rightarrow 2\,CaO$

(ii) Is the entropy change for this reaction positive or negative? Briefly explain.

(ii) Entropy is decreasing during this reaction because 3 moles of reactants combine to make 2 moles of products and because a solid reacts with a gas to form a solid.

(c) Ammonia and boron trichloride gases are mixed.

(i) Balanced equation:

(i) $NH_3 + BCl_3 \Rightarrow H_3NBCl_3$

(ii) Identify the Lewis acid in this reaction.

(ii) Boron trichloride accepts the electron pair from ammonia, so BCl_3 acts as the Lewis acid.

QUESTIONS	EXPLANATIONS

5. Use your knowledge of chemical principles to answer the following questions.

(a) The table above shows the first and second ionization energies for potassium and sodium

    (i) Explain the difference between the first ionization energies of potassium and sodium.

        (i) The first ionization energy is lower for potassium than for sodium. Ionization energy is the energy required for the removal of an electron. Potassium's single valence electron is located in the fourth shell, while sodium's single valence electron is in the third shell. So potassium's electron is farther from the nucleus and shielded by an extra shell, making it easier to remove.

    (ii) Explain why the second ionization energy for potassium is so much larger than the first ionization energy.

        (ii) Potassium's first electron is the only valence electron in the fourth shell, so it is easy to remove. The second electron must be removed from a completed, stable low energy third shell, which is much more difficult.

(b) The table above shows the atomic radii of several atoms and ions.

    (i) The atomic radius of bromine is smaller than the atomic radius of potassium. Explain.

        (i) Potassium and bromine each have four shells, but bromine has many more protons in its nucleus than does potassium. Bromine's extra protons exert greater attractive force on its electrons, pulling them closer to the nucleus and making the atomic radius smaller.

    (ii) The potassium and chlorine ions have the same electronic structure. Explain the difference in their atomic radii.

        (ii) Both K^+ and Cl^- have the same electron configuration, that of neutral argon. But K^+ has two more protons in its nucleus. The extra protons exert greater attractive force on the potassium ion's electrons, pulling them closer to the nucleus and making the ionic radius smaller.

    (iii) When potassium bromide and sodium chloride are compared, which will have the higher melting point? Explain.

        (iii) KBr and NaCl solids are held together by ionic bonds. Ionic bond strength depends on two things: the size of the charges and the distance between them. All of the charges in KBr and NaCl are +1 or −1, so the difference in melting points must be based on distance. NaCl is composed of smaller ions, so the charges will be closer together, making the electrostatic attractions greater, the bonds stronger, and the melting point higher.

(c) The table above shows the boiling points for several organic compounds. Explain the trend in increasing boiling point.

    (c) Methane (CH_4) is nonpolar and its molecules have only weak van der Waals attractions, so it has the lowest boiling point. Methoxymethane (CH_3OCH_3) is larger and slightly polar, so its stronger intermolecular attractions make for a higher boiling point. Ethanol (CH_3CH_2OH) exhibits hydrogen bonding, so it has the strongest intermolecular forces and the highest boiling point of the three.

QUESTIONS	EXPLANATIONS

6. An experiment is to be performed involving the titration of a solution of benzoic acid, $HC_7H_5O_2$, of unknown concentration with a standardized 1.00-molar NaOH solution. The pK_a of benzoic acid is 4.2.

(a) What measurements and calculations must be made if the concentration of the acid is to be determined?

(a) The volume of the acid must be measured before the titration begins. Then the volume of 1.00-molar NaOH solution required to reach the equivalence point must be measured.

Once the two volumes are known, the concentration of the acid can be found with the following calculation:

$$(M_{NaOH})(V_{NaOH\ added}) = (M_{acid})(V_{acid})$$

(b) An indicator must be chosen for the titration. If methyl red (pH interval for color change: 4.2–6.3) and thymol blue (pH interval for color change: 8.0–9.6) are the only indicators available, which one should be used? Explain your choice.

(b) Thymol blue should be chosen. When a weak acid reacts with a strong base, the equivalence point will be greater than 7 in the basic region. The color change for thymol blue occurs in the basic region, making it the more appropriate choice for this experiment.

(c) Describe how you could use the titration apparatus and benzoic acid and NaOH solutions to create a buffer solution with a pH of 4.2 after you measured the volume of NaOH solution required to reach the equivalence point for a given volume of benzoic acid solution.

(c) Start the titration over again with a fresh sample of benzoic acid solution of the same volume that was used to determine the equivalence point. Add exactly half the volume of NaOH solution this time. At this point, called the half-equivalence point, exactly half of the benzoic acid has been neutralized and the concentration of benzoic acid is equal to the concentration of its conjugate base, benzoate ion. At this point, $pH = pK_a$.

(d) Describe a way to test whether the neutralization of benzoic acid by a strong base is an exothermic process, using readily available school laboratory equipment.

(d) Put a thermometer in the calorimeter used for the titration. Measure the temperature of the acid solution before the titration. If the temperature increases during the titration, then energy is being released and the reaction is exothermic.

QUESTIONS	EXPLANATIONS

7.

$$H_3PO_4 \rightleftharpoons H^+ + H_2PO_4^- \qquad K_1 = 7.5 \times 10^{-3}$$

$$H_2PO_4^- \rightleftharpoons H^+ + HPO_4^{2-} \qquad K_2 = 6.2 \times 10^{-8}$$

$$HPO_4^{2-} \rightleftharpoons H^+ + PO_4^{3-} \qquad K_3 = 2.2 \times 10^{-13}$$

(a) Choose an amphoteric species from the reactions listed above, and give its conjugate acid and its conjugate base.

(a) An amphoteric species can act either as an acid or base.

$H_2PO_4^-$ is amphoteric.

Conjugate base: HPO_4^{2-}

Conjugate acid: H_3PO_4

or

HPO_4^{2-} is amphoteric.

Conjugate base: PO_4^{3-}

Conjugate acid: $H_2PO_4^-$

(b) Explain why the dissociation constant decreases with each hydrogen ion lost.

(b) The smaller the dissociation constant, the more tightly the acid holds its hydrogen ions and the weaker the acid becomes.

After a hydrogen ion has been removed, the remaining species has a negative charge that attracts the remaining hydrogen ions more strongly.

(c) Of the acids listed above, which would be most useful in creating a buffer solution with a pH of 7.5?

(c) $$H_2PO_4^- \rightleftharpoons H^+ + HPO_4^{2-} \qquad K_2 = 6.2 \times 10^{-8}$$

For $H_2PO_4^-$, $pK_a = -\log(6.2 \times 10^{-8}) = 7.2$

The best buffer solution is made with an acid whose pK_a is approximately equal to the desired pH.

QUESTIONS	EXPLANATIONS

(d) Sketch the titration curve that results when H_3PO_4 is titrated with excess NaOH and label the two axes.

(d)

The axes should have the correct labeling (pH and NaOH added).

The curve should show rising pH, and it should have three bumps.

ABOUT THE AUTHOR

Paul Foglino has taught for The Princeton Review for more than a decade. He has written and edited course materials for the SAT, GRE, GMAT, and MCAT courses. He is author of *Math Smart Jr. II* and coauthor of *Cracking the CLEP*. Foglino studied English and electrical engineering at Columbia University, but he remains convinced that he learned everything he ever needed to know in junior high school.

Completely darken bubbles with a No. 2 pencil. If you make a mistake, be sure to erase mark completely. Erase all stray marks.

1.

YOUR NAME: _____
(Print)
Last   First   M.I.

SIGNATURE: _____   DATE: ___ / ___ / ___

HOME ADDRESS: _____
(Print)
Number and Street

_____
City   State   Zip Code

PHONE NO.: _____

5. YOUR NAME

First 4 letters of last name				FIRST INIT	MID INIT
Ⓐ	Ⓐ	Ⓐ	Ⓐ	Ⓐ	Ⓐ
Ⓑ	Ⓑ	Ⓑ	Ⓑ	Ⓑ	Ⓑ
Ⓒ	Ⓒ	Ⓒ	Ⓒ	Ⓒ	Ⓒ
Ⓓ	Ⓓ	Ⓓ	Ⓓ	Ⓓ	Ⓓ
Ⓔ	Ⓔ	Ⓔ	Ⓔ	Ⓔ	Ⓔ
Ⓕ	Ⓕ	Ⓕ	Ⓕ	Ⓕ	Ⓕ
Ⓖ	Ⓖ	Ⓖ	Ⓖ	Ⓖ	Ⓖ
Ⓗ	Ⓗ	Ⓗ	Ⓗ	Ⓗ	Ⓗ
Ⓘ	Ⓘ	Ⓘ	Ⓘ	Ⓘ	Ⓘ
Ⓙ	Ⓙ	Ⓙ	Ⓙ	Ⓙ	Ⓙ
Ⓚ	Ⓚ	Ⓚ	Ⓚ	Ⓚ	Ⓚ
Ⓛ	Ⓛ	Ⓛ	Ⓛ	Ⓛ	Ⓛ
Ⓜ	Ⓜ	Ⓜ	Ⓜ	Ⓜ	Ⓜ
Ⓝ	Ⓝ	Ⓝ	Ⓝ	Ⓝ	Ⓝ
Ⓞ	Ⓞ	Ⓞ	Ⓞ	Ⓞ	Ⓞ
Ⓟ	Ⓟ	Ⓟ	Ⓟ	Ⓟ	Ⓟ
Ⓠ	Ⓠ	Ⓠ	Ⓠ	Ⓠ	Ⓠ
Ⓡ	Ⓡ	Ⓡ	Ⓡ	Ⓡ	Ⓡ
Ⓢ	Ⓢ	Ⓢ	Ⓢ	Ⓢ	Ⓢ
Ⓣ	Ⓣ	Ⓣ	Ⓣ	Ⓣ	Ⓣ
Ⓤ	Ⓤ	Ⓤ	Ⓤ	Ⓤ	Ⓤ
Ⓥ	Ⓥ	Ⓥ	Ⓥ	Ⓥ	Ⓥ
Ⓦ	Ⓦ	Ⓦ	Ⓦ	Ⓦ	Ⓦ
Ⓧ	Ⓧ	Ⓧ	Ⓧ	Ⓧ	Ⓧ
Ⓨ	Ⓨ	Ⓨ	Ⓨ	Ⓨ	Ⓨ
Ⓩ	Ⓩ	Ⓩ	Ⓩ	Ⓩ	Ⓩ

IMPORTANT: Please fill in these boxes exactly as shown on the back cover of your test book.

2. TEST FORM

_____

3. TEST CODE

⓪	Ⓐ	Ⓙ	⓪	⓪
①	Ⓑ	Ⓚ	①	①
②	Ⓒ	Ⓛ	②	②
③	Ⓓ	Ⓜ	③	③
④	Ⓔ	Ⓝ	④	④
⑤	Ⓕ	Ⓞ	⑤	⑤
⑥	Ⓖ	Ⓟ	⑥	⑥
⑦	Ⓗ	Ⓠ	⑦	⑦
⑧	Ⓘ	Ⓡ	⑧	⑧
⑨			⑨	⑨

4. REGISTRATION NUMBER

⓪	⓪	⓪	⓪	⓪	⓪	⓪
①	①	①	①	①	①	①
②	②	②	②	②	②	②
③	③	③	③	③	③	③
④	④	④	④	④	④	④
⑤	⑤	⑤	⑤	⑤	⑤	⑤
⑥	⑥	⑥	⑥	⑥	⑥	⑥
⑦	⑦	⑦	⑦	⑦	⑦	⑦
⑧	⑧	⑧	⑧	⑧	⑧	⑧
⑨	⑨	⑨	⑨	⑨	⑨	⑨

6. DATE OF BIRTH

Month		Day		Year	
◯ JAN					
◯ FEB	⓪	⓪	⓪	⓪	
◯ MAR	①	①	①	①	
◯ APR	②	②	②	②	
◯ MAY	③	③	③	③	
◯ JUN		④	④	④	
◯ JUL		⑤	⑤	⑤	
◯ AUG		⑥	⑥	⑥	
◯ SEP		⑦	⑦	⑦	
◯ OCT		⑧	⑧	⑧	
◯ NOV		⑨	⑨	⑨	
◯ DEC					

7. GENDER
◯ MALE
◯ FEMALE

The Princeton Review®

1. Ⓐ Ⓑ Ⓒ Ⓓ
2. Ⓐ Ⓑ Ⓒ Ⓓ
3. Ⓐ Ⓑ Ⓒ Ⓓ
4. Ⓐ Ⓑ Ⓒ Ⓓ
5. Ⓐ Ⓑ Ⓒ Ⓓ
6. Ⓐ Ⓑ Ⓒ Ⓓ
7. Ⓐ Ⓑ Ⓒ Ⓓ
8. Ⓐ Ⓑ Ⓒ Ⓓ
9. Ⓐ Ⓑ Ⓒ Ⓓ
10. Ⓐ Ⓑ Ⓒ Ⓓ
11. Ⓐ Ⓑ Ⓒ Ⓓ
12. Ⓐ Ⓑ Ⓒ Ⓓ
13. Ⓐ Ⓑ Ⓒ Ⓓ
14. Ⓐ Ⓑ Ⓒ Ⓓ
15. Ⓐ Ⓑ Ⓒ Ⓓ
16. Ⓐ Ⓑ Ⓒ Ⓓ
17. Ⓐ Ⓑ Ⓒ Ⓓ
18. Ⓐ Ⓑ Ⓒ Ⓓ
19. Ⓐ Ⓑ Ⓒ Ⓓ
20. Ⓐ Ⓑ Ⓒ Ⓓ

21. Ⓐ Ⓑ Ⓒ Ⓓ
22. Ⓐ Ⓑ Ⓒ Ⓓ
23. Ⓐ Ⓑ Ⓒ Ⓓ
24. Ⓐ Ⓑ Ⓒ Ⓓ
25. Ⓐ Ⓑ Ⓒ Ⓓ
26. Ⓐ Ⓑ Ⓒ Ⓓ
27. Ⓐ Ⓑ Ⓒ Ⓓ
28. Ⓐ Ⓑ Ⓒ Ⓓ
29. Ⓐ Ⓑ Ⓒ Ⓓ
30. Ⓐ Ⓑ Ⓒ Ⓓ
31. Ⓐ Ⓑ Ⓒ Ⓓ
32. Ⓐ Ⓑ Ⓒ Ⓓ
33. Ⓐ Ⓑ Ⓒ Ⓓ
34. Ⓐ Ⓑ Ⓒ Ⓓ
35. Ⓐ Ⓑ Ⓒ Ⓓ
36. Ⓐ Ⓑ Ⓒ Ⓓ
37. Ⓐ Ⓑ Ⓒ Ⓓ
38. Ⓐ Ⓑ Ⓒ Ⓓ
39. Ⓐ Ⓑ Ⓒ Ⓓ
40. Ⓐ Ⓑ Ⓒ Ⓓ

41. Ⓐ Ⓑ Ⓒ Ⓓ
42. Ⓐ Ⓑ Ⓒ Ⓓ
43. Ⓐ Ⓑ Ⓒ Ⓓ
44. Ⓐ Ⓑ Ⓒ Ⓓ
45. Ⓐ Ⓑ Ⓒ Ⓓ
46. Ⓐ Ⓑ Ⓒ Ⓓ
47. Ⓐ Ⓑ Ⓒ Ⓓ
48. Ⓐ Ⓑ Ⓒ Ⓓ
49. Ⓐ Ⓑ Ⓒ Ⓓ
50. Ⓐ Ⓑ Ⓒ Ⓓ
51. Ⓐ Ⓑ Ⓒ Ⓓ
52. Ⓐ Ⓑ Ⓒ Ⓓ
53. Ⓐ Ⓑ Ⓒ Ⓓ
54. Ⓐ Ⓑ Ⓒ Ⓓ
55. Ⓐ Ⓑ Ⓒ Ⓓ
56. Ⓐ Ⓑ Ⓒ Ⓓ
57. Ⓐ Ⓑ Ⓒ Ⓓ
58. Ⓐ Ⓑ Ⓒ Ⓓ
59. Ⓐ Ⓑ Ⓒ Ⓓ
60. Ⓐ Ⓑ Ⓒ Ⓓ

The Princeton Review

Completely darken bubbles with a No. 2 pencil. If you make a mistake, be sure to erase mark completely. Erase all stray marks.

1.

YOUR NAME: _____
(Print)   Last   First   M.I.

SIGNATURE: _____   DATE: __ / __ / __

HOME ADDRESS: _____
(Print)   Number and Street

_____
City   State   Zip Code

PHONE NO.: _____

IMPORTANT: Please fill in these boxes exactly as shown on the back cover of your test book.

2. TEST FORM

_____

3. TEST CODE

0	Ⓐ	Ⓙ	0	0
1	Ⓑ	Ⓚ	1	1
2	Ⓒ	Ⓛ	2	2
3	Ⓓ	Ⓜ	3	3
4	Ⓔ	Ⓝ	4	4
5	Ⓕ	Ⓞ	5	5
6	Ⓖ	Ⓟ	6	6
7	Ⓗ	Ⓠ	7	7
8	Ⓘ	Ⓡ	8	8
9			9	9

4. REGISTRATION NUMBER

0	0	0	0	0	0	0
1	1	1	1	1	1	1
2	2	2	2	2	2	2
3	3	3	3	3	3	3
4	4	4	4	4	4	4
5	5	5	5	5	5	5
6	6	6	6	6	6	6
7	7	7	7	7	7	7
8	8	8	8	8	8	8
9	9	9	9	9	9	9

5. YOUR NAME

First 4 letters of last name				FIRST INIT	MID INIT
Ⓐ	Ⓐ	Ⓐ	Ⓐ	Ⓐ	Ⓐ
Ⓑ	Ⓑ	Ⓑ	Ⓑ	Ⓑ	Ⓑ
Ⓒ	Ⓒ	Ⓒ	Ⓒ	Ⓒ	Ⓒ
Ⓓ	Ⓓ	Ⓓ	Ⓓ	Ⓓ	Ⓓ
Ⓔ	Ⓔ	Ⓔ	Ⓔ	Ⓔ	Ⓔ
Ⓕ	Ⓕ	Ⓕ	Ⓕ	Ⓕ	Ⓕ
Ⓖ	Ⓖ	Ⓖ	Ⓖ	Ⓖ	Ⓖ
Ⓗ	Ⓗ	Ⓗ	Ⓗ	Ⓗ	Ⓗ
Ⓘ	Ⓘ	Ⓘ	Ⓘ	Ⓘ	Ⓘ
Ⓙ	Ⓙ	Ⓙ	Ⓙ	Ⓙ	Ⓙ
Ⓚ	Ⓚ	Ⓚ	Ⓚ	Ⓚ	Ⓚ
Ⓛ	Ⓛ	Ⓛ	Ⓛ	Ⓛ	Ⓛ
Ⓜ	Ⓜ	Ⓜ	Ⓜ	Ⓜ	Ⓜ
Ⓝ	Ⓝ	Ⓝ	Ⓝ	Ⓝ	Ⓝ
Ⓞ	Ⓞ	Ⓞ	Ⓞ	Ⓞ	Ⓞ
Ⓟ	Ⓟ	Ⓟ	Ⓟ	Ⓟ	Ⓟ
Ⓠ	Ⓠ	Ⓠ	Ⓠ	Ⓠ	Ⓠ
Ⓡ	Ⓡ	Ⓡ	Ⓡ	Ⓡ	Ⓡ
Ⓢ	Ⓢ	Ⓢ	Ⓢ	Ⓢ	Ⓢ
Ⓣ	Ⓣ	Ⓣ	Ⓣ	Ⓣ	Ⓣ
Ⓤ	Ⓤ	Ⓤ	Ⓤ	Ⓤ	Ⓤ
Ⓥ	Ⓥ	Ⓥ	Ⓥ	Ⓥ	Ⓥ
Ⓦ	Ⓦ	Ⓦ	Ⓦ	Ⓦ	Ⓦ
Ⓧ	Ⓧ	Ⓧ	Ⓧ	Ⓧ	Ⓧ
Ⓨ	Ⓨ	Ⓨ	Ⓨ	Ⓨ	Ⓨ
Ⓩ	Ⓩ	Ⓩ	Ⓩ	Ⓩ	Ⓩ

6. DATE OF BIRTH

Month		Day		Year	
◯ JAN					
◯ FEB	0	0	0	0	
◯ MAR	1	1	1	1	
◯ APR	2	2	2	2	
◯ MAY	3	3	3	3	
◯ JUN		4	4	4	
◯ JUL		5	5	5	
◯ AUG		6	6	6	
◯ SEP		7	7	7	
◯ OCT		8	8	8	
◯ NOV		9	9	9	
◯ DEC					

7. GENDER
◯ MALE
◯ FEMALE

The Princeton Review

1. Ⓐ Ⓑ Ⓒ Ⓓ
2. Ⓐ Ⓑ Ⓒ Ⓓ
3. Ⓐ Ⓑ Ⓒ Ⓓ
4. Ⓐ Ⓑ Ⓒ Ⓓ
5. Ⓐ Ⓑ Ⓒ Ⓓ
6. Ⓐ Ⓑ Ⓒ Ⓓ
7. Ⓐ Ⓑ Ⓒ Ⓓ
8. Ⓐ Ⓑ Ⓒ Ⓓ
9. Ⓐ Ⓑ Ⓒ Ⓓ
10. Ⓐ Ⓑ Ⓒ Ⓓ
11. Ⓐ Ⓑ Ⓒ Ⓓ
12. Ⓐ Ⓑ Ⓒ Ⓓ
13. Ⓐ Ⓑ Ⓒ Ⓓ
14. Ⓐ Ⓑ Ⓒ Ⓓ
15. Ⓐ Ⓑ Ⓒ Ⓓ
16. Ⓐ Ⓑ Ⓒ Ⓓ
17. Ⓐ Ⓑ Ⓒ Ⓓ
18. Ⓐ Ⓑ Ⓒ Ⓓ
19. Ⓐ Ⓑ Ⓒ Ⓓ
20. Ⓐ Ⓑ Ⓒ Ⓓ

21. Ⓐ Ⓑ Ⓒ Ⓓ
22. Ⓐ Ⓑ Ⓒ Ⓓ
23. Ⓐ Ⓑ Ⓒ Ⓓ
24. Ⓐ Ⓑ Ⓒ Ⓓ
25. Ⓐ Ⓑ Ⓒ Ⓓ
26. Ⓐ Ⓑ Ⓒ Ⓓ
27. Ⓐ Ⓑ Ⓒ Ⓓ
28. Ⓐ Ⓑ Ⓒ Ⓓ
29. Ⓐ Ⓑ Ⓒ Ⓓ
30. Ⓐ Ⓑ Ⓒ Ⓓ
31. Ⓐ Ⓑ Ⓒ Ⓓ
32. Ⓐ Ⓑ Ⓒ Ⓓ
33. Ⓐ Ⓑ Ⓒ Ⓓ
34. Ⓐ Ⓑ Ⓒ Ⓓ
35. Ⓐ Ⓑ Ⓒ Ⓓ
36. Ⓐ Ⓑ Ⓒ Ⓓ
37. Ⓐ Ⓑ Ⓒ Ⓓ
38. Ⓐ Ⓑ Ⓒ Ⓓ
39. Ⓐ Ⓑ Ⓒ Ⓓ
40. Ⓐ Ⓑ Ⓒ Ⓓ

41. Ⓐ Ⓑ Ⓒ Ⓓ
42. Ⓐ Ⓑ Ⓒ Ⓓ
43. Ⓐ Ⓑ Ⓒ Ⓓ
44. Ⓐ Ⓑ Ⓒ Ⓓ
45. Ⓐ Ⓑ Ⓒ Ⓓ
46. Ⓐ Ⓑ Ⓒ Ⓓ
47. Ⓐ Ⓑ Ⓒ Ⓓ
48. Ⓐ Ⓑ Ⓒ Ⓓ
49. Ⓐ Ⓑ Ⓒ Ⓓ
50. Ⓐ Ⓑ Ⓒ Ⓓ
51. Ⓐ Ⓑ Ⓒ Ⓓ
52. Ⓐ Ⓑ Ⓒ Ⓓ
53. Ⓐ Ⓑ Ⓒ Ⓓ
54. Ⓐ Ⓑ Ⓒ Ⓓ
55. Ⓐ Ⓑ Ⓒ Ⓓ
56. Ⓐ Ⓑ Ⓒ Ⓓ
57. Ⓐ Ⓑ Ⓒ Ⓓ
58. Ⓐ Ⓑ Ⓒ Ⓓ
59. Ⓐ Ⓑ Ⓒ Ⓓ
60. Ⓐ Ⓑ Ⓒ Ⓓ

NOTES

NOTES